Digital Transformation and Enterprise Information Systems

Today, only having information is insufficient for a competitive advantage. To be ahead of your competitors, you need to know them, and for this, it is important to constantly manage news and content. The information needs to be condensed and analyzed so that it can be useful to decision-makers. Enterprise systems let you collect, organize, analyze, and disseminate content so it can serve your business goals quickly. It is a daily business process, particularly for small businesses dealing with large amounts of information and digital transformation process. How you handle this information can be a competitive advantage for businesses.

Digital Transformation and Enterprise Information Systems introduces new ways to use new digital technologies to enable major business improvements, enhance customer experience, simplify operations, and create new business models. It provides a better understanding of the impact of digital transformation on how enterprise systems and enterprise information technology services are managed and delivered. The book offers new knowledge on how enterprise systems influence digital business transformation and its role, along with addressing several dimensions of digital transformation in favor of SMEs. The result provides business leaders with tools and strategies to manage their enterprises better.

This reference book can be used by researchers, undergraduate students taking engineering and management courses, and managers, engineers, and other professionals in fields related to technologies, management, and engineering innovation.

Manufacturing Design and Technology Series

Series Editor: J. Paulo Davim, University of Aveiro, Portugal

This series will publish high quality references and advanced textbooks in the broad area of manufacturing design and technology, with a special focus on sustainability in manufacturing. Books in the series should find a balance between academic research and industrial application. This series targets academics and practicing engineers working on topics in materials science, mechanical engineering, industrial engineering, systems engineering, and environmental engineering as related to manufacturing systems, as well as professions in manufacturing design.

Industry 4.0: Challenges, Trends, and Solutions in Management and Engineering
Edited by Carolina Machado and J. Paulo Davim

Sustainable Manufacturing for Industry 4.0
An Augmented Approach
Edited by K. Jayakrishna, K.E.K. Vimal, S. Aravind Raj, Asela K. Kulatunga, M.T.H. Sultan, and J. Paulo Davim

Functional and Smart Materials
Edited by Chander Prakash, Sunpreet Singh, and J. Paulo Davim

Advanced Manufacturing and Processing Technology
Edited by Chander Prakash, Sunpreet Singh, and J. Paulo Davim

Characterization, Testing, Measurement, and Metrology
Edited by Chander Prakash, Sunpreet Singh, and J. Paulo Davim

Non-Conventional Hybrid Machining Processes
Theory and Practice
Edited by Rupinder Singh and J. Paulo Davim

Digital Transformation and Enterprise Information Systems
Edited by Adelaide Martins and Carolina Machado

For more information about this series, please visit: https://www.crcpress.com/Manufacturing-Design-and-Technology/book-series/CRCMANDESTEC

Digital Transformation and Enterprise Information Systems

Edited by
Adelaide Martins and Carolina Machado

CRC Press
Taylor & Francis Group
Boca Raton London New York

CRC Press is an imprint of the
Taylor & Francis Group, an **informa** business

Designed cover image: Shutterstock – metamorworks

First edition published 2025
by CRC Press
2385 NW Executive Center Drive, Suite 320, Boca Raton FL 33431

and by CRC Press
4 Park Square, Milton Park, Abingdon, Oxon, OX14 4RN

CRC Press is an imprint of Taylor & Francis Group, LLC

ISBN: 978-1-032-61000-9 (hbk)
ISBN: 978-1-032-61065-8 (pbk)
ISBN: 978-1-003-46182-1 (ebk)

DOI: 10.1201/9781003461821

Typeset in Times
by SPi Technologies India Pvt Ltd (Straive)

Contents

Preface

The crisis with the outbreak of Covid-19, the extreme weather phenomena, and the recent war in Europe have revealed that disruptive contexts will be increasingly recurrent in the future. To operate under conditions of deep uncertainty, all types of organizations, with a particular emphasis on SMEs, need to keep up with technological changes and embark on a digital transformation. Through business digital transformation, they can make better data-driven decisions tailored to their own business needs. One of the ways for any firm to start its digital business transformation journey is through the implementation of enterprise systems. Digital transformation is triggering significant changes in firms' business processes and in their enterprise information systems, such as enterprise resource planning, supply chain management, customer relationship management, and enterprise social platforms, among others. Enterprise systems can be used to streamline and integrate an organization's business processes and to significantly improve levels of information and knowledge within the organization as well as with its customers and suppliers. New digital systems such as wireless sensors embedded in products, analytics tools, and artificial intelligence are also impacting enterprise systems and creating new organizational challenges. This field of research still being emergent, both in big organizations and in the SME context, with this book we look to highlight the latest and more recent research findings that are occurring in this field. In particular, this book looks to: introduce new lines of research in the use of new digital technologies to enable major business improvements, enhance customer experience, simplify operations, or create new business models; provide a better understanding of the impact of digital transformation on how enterprise systems and enterprise IT services are managed and delivered; and provide new knowledge on how enterprise systems influence digital business transformation.

For this purpose, this book, entitled *Digital Transformation and Enterprise Information Systems*, is divided into eight chapters: Chapter 1 covers "Robotic Process Automation: Transforming Enterprise Processes"; Chapter 2 describes "Enterprise Resource Planning and Business Intelligence Integration and Management Control Change: Managing Tensions and Hybridity"; Chapter 3 discusses "Firm Internationalization and Activity Impact on Digital Technologies Adoption"; Chapter 4 describes "Recommendation Systems in E-commerce: Link Prediction in Multilayer Bipartite Networks"; Chapter 5 addresses "Digital Transformation and Business Processes: An Experience-Based Perspective"; Chapter 6 covers "Improving Decision Making with Business Intelligence: A Case Study of Lusilectra"; Chapter 7 describes "Diagnosing the Implementation of Digital Technologies by European Union Manufacturing Multinationals"; and Chapter 8 provides "A Framework for Enabling Effective Digital Transformation in Organizations Based in Business Processes."

This book can be used as a research book for final undergraduate engineering and management courses (e.g., management, industrial engineering, international business, production engineering, etc.) or as a subject on management and engineering innovation at the postgraduate level. Also, it can serve as a useful reference for academics, researchers, managers, engineers, and other professionals in fields related to technologies, management, and engineering innovation.

The editors are grateful to CRC Press/Taylor & Francis Group for this opportunity and their professional support. Finally, we thank all chapter contributors for their interest and the time allotted to work on this project.

Adelaide Martins
Porto, Portugal

Carolina Machado
Braga, Portugal

Contributors

Pedro Campos
University of Porto, School of Economics and Management and Laboratory of Artificial Intelligence and Decision Support – INESC TEC
Porto, Portugal

Vítor Correia
Lusilectra, Salvador Caetano Group
Porto, Portugal

Bárbara Corvacho
University of Porto, School of Economics and Management
Porto, Portugal

Francisco J. Duarte
Engineering School, University of Minho, and CCG/ZGDV Innovation Institute
Guimarães, Portugal

Daniel Freitas
University of Porto, School of Economics and Management
Porto, Portugal

Sofia Gomes
University Portucalense and REMIT – Research on Economics, Management and Information Technologies
Porto, Portugal

João Pedro Ladeiras
Deloitte Portugal
Porto, Portugal

João M. Lopes
Miguel Torga Institute of Higher Education, Coimbra, Portugal & NECE-UBI – Research Unit in Business Sciences, University of Beira Interior
Covilhã, Portugal

Carolina Machado
School of Economics and Management, University of Minho
and
Interdisciplinary Centre of Social Sciences (CICS.NOVA.UMinho)
Braga, Portugal

Adelaide Martins
University of Porto, School of Economics and Management, and CEF.UP
Porto, Portugal

João Oliveira
University of Porto, School of Economics and Management
Porto, Portugal

Luís Ramoa
University of Porto, School of Economics and Management
Porto, Portugal

Carlos Ribeiro
University Portucalense and REMIT – Research on Economics, Management and Information Technologies
Porto, Portugal

About the Editors

Adelaide Martins is an Assistant Professor at the School of Economics and Management of the University of Porto, Portugal. She has a PhD degree in Business Administration and a Master's degree in Industrial and Business Economics, both from the University of Minho. Her teaching interests are in accounting and management information systems. Her current main research projects are: Accounting Education; Social and Environmental Reporting; Accounting and Auditing in the Era of Digital Transformation. She has published several book chapters and articles in academic journals such as *Critical Perspectives on Accounting*, *Meditari Accountancy Research*, *Qualitative Research in Accounting and Management*, *Journal of International Accounting, Auditing & Taxation*, *Accounting Education and Utilities Policy*, among others.

Carolina Machado received her PhD degree in Management Sciences (Organizational and Politics Management area/Human Resources Management) from the University of Minho in 1999, Master's degree in Management (Strategic Human Resource Management) from Technical University of Lisbon in 1994, and Degree in Business Administration from the University of Minho in 1989. Teaching in the Human Resources Management subjects since 1989 at the University of Minho, she is since 2004 an Associated Professor (with Habilitation since May 2022), with experience and research interest areas in the fields of Human Resource Management, International Human Resource Management, Human Resource Management in SMEs, Training and Development, Emotional Intelligence, Management Change, Knowledge Management and Management/HRM in the Digital Age/Business Analytics. She is Head of the Human Resources Management Work Group at the School of Economics and Management at University of Minho, Coordinator of Advanced Training Courses at the Interdisciplinary Centre of Social Sciences, Member of the Interdisciplinary Centre of Social Sciences (CICS.NOVA.UMinho), University of Minho, and Chief Editor of the *International Journal of Applied Management Sciences and Engineering* (*IJAMSE*), guest editor of journals, book editor and book series editor, and reviewer in different international prestigious journals. In addition, she has also published both as editor/co-editor and as author/co-author several books, book chapters, and articles in journals and conferences.

1 Robotic Process Automation

Transforming Enterprise Processes

João Pedro Ladeiras
Deloitte Portugal, Porto, Portugal

Adelaide Martins
University of Porto, Porto, Portugal

1.1 INTRODUCTION

In today's challenging business environment, companies find themselves navigating a terrain characterized by relentless industry competition, punctuated by disruptive events that have the potential to rapidly reshape business landscapes (Bundy et al., 2017; Hällgren et al., 2018; Sharma et al., 2020; Winston, 2020). The dynamics of the business world have undergone a metamorphosis, driven by both internal and external forces.

In the 21st century, competitiveness has reached unprecedented levels, with industries witnessing the emergence of fierce rivalries and market battles. A notable example is the technology industry, where giants such as Amazon, Microsoft, and Google are engaged in a constant race for innovation and market dominance. The battle for cloud computing supremacy between Amazon Web Services (AWS), Microsoft Azure, and Google Cloud is a clear illustration of the intense competition in the sector.

This relentless competition sets the backdrop for a changing business narrative, where once an occasional disruptor has become a continuous disruptor. The COVID-19 pandemic, a seismic event of unprecedented proportions, exposed the vulnerabilities of the global supply chains, business continuity plans, and the traditional office-centric work model (Sharma, 2023; Winston, 2020). Industries such as travel, hospitality, and retail faced existential threats, forcing them to reinvent business models on the fly (Donthu & Gustafsson, 2020; Thakur & Hale, 2022). The need for resilience and adaptability became clear as companies scrambled to embrace remote working, accelerate digital transformations, and fortify their operations against future setbacks.

The abrupt geopolitical shifts of recent years have further exacerbated the challenges faced by companies operating on the global stage (Ahlstrom et al., 2020). Trade tensions, sanctions, and political uncertainties have added a layer of complexity

DOI: 10.1201/9781003461821-1

to strategic decision-making processes. Escalating trade tensions between the U.S. and China have forced multinational corporations to reassess their supply chain strategies, diversify their sourcing, and navigate a landscape of geopolitical uncertainties. Additionally, conflicts such as the Ukrainian–Russian war and other global geopolitical developments underscore the volatile nature of international relations, pushing companies to adopt agile strategies that account for a dynamic geopolitical landscape.

Technology, as the basis of contemporary business strategies, plays a fundamental role in enabling companies to thrive amid uncertainty (Koch & Windsperger, 2017; Soto-Acosta, 2024). From automating routine tasks to providing insights through data analytics, technology empowers organizations to navigate the complexities of the business landscape. Within this technological toolkit, Robotic Process Automation (RPA) emerges as a key player (Deloitte, 2023; Enriquez et al., 2020; Huang & Vasarhelyi, 2019), offering a nuanced and efficient response to the challenges posed by industry competition and disruptive events.

In this chapter, we aim to frame important issues for the industry dynamics emerging from the strategic adoption of RPA. As we explore the significance of RPA in this adaptive landscape, we uncover its applications, benefits, and transformative role in reshaping how organizations approach operational challenges. We expect that this chapter will motivate a fruitful debate among undergraduate and postgraduate students, academics, managers, and other professionals concerned with how best to respond to the opportunities and challenges of RPA.

The chapter is structured as follows: Section 1.2 provides an overview of RPA and how this technology operates. Section 1.3 describes how RPA can support organizations with cost reduction and increase in efficiency, Section 1.4 describes the impact of the RPA on human capital, and Section 1.5 delves into the main challenges, limitations, and future trends associated with the use of RPA. Section 1.6 provides the concluding remarks.

1.2 UNDERSTANDING ROBOTIC PROCESS AUTOMATION

1.2.1 RPA Relevance to Contemporary Business Environments

Exploring the details of RPA requires establishing a solid foundation by comprehensively understanding the core elements that define this transformative technology. RPA is an advanced technology that utilizes software "robots" to emulate and automate human interactions with digital systems, representing a paradigm shift in how organizations approach routine, rule-driven tasks (Enriquez et al., 2020). Armed with the ability to navigate user interfaces and manipulate data, these software robots perform tasks traditionally performed by humans (Deloitte, 2017). This includes, but is not limited to, data entry, information retrieval, and rule-based decision-making processes.

Fundamentally, RPA operates on the principle of rule-based automation, where pre-defined logical rules govern the actions of the software robots. These rules serve as a blueprint for automating workflows, ensuring accuracy, consistency, and efficiency in the execution of tasks. As we navigate this landscape, it is essential to recognize the nuanced interplay between RPA and existing enterprise systems that form the backbone of streamlined and automated business processes.

RPA has distinctive characteristics that set it apart as a revolutionary force in business process optimization. First and foremost, its agility – RPA allows organizations to swiftly adapt to changing business requirements without the need for extensive coding or system overhauls. The user-friendly interface enables quick deployment, reducing the time to value and enhancing operational efficiency. The non-invasive nature of RPA is another hallmark characteristic. Unlike traditional methods, which may require extensive modifications to existing systems, RPA integrates seamlessly with the prevailing Information Technology (IT) infrastructure. This adaptability facilitates a phased approach to automation, allowing organizations to reap the benefits without disrupting established workflows. RPA also brings unprecedented accuracy to mundane, repetitive tasks, mitigating the risk of human error. It operates tirelessly, 24/7, enhancing productivity and freeing up human resources to focus on more strategic, value-added activities (Deloitte, 2017). The cost-effectiveness of implementing RPA, compared to the labor-intensive nature of traditional methods, is a compelling economic driver and an additional valuable advantage.

In the digital era, where the pace of change is accelerating and data is the lifeblood of organizations, RPA is emerging as a fundamental technology in digital transformation strategies. Its role transcends mere task automation; RPA is becoming an enabler of agility, innovation, and scalability. By automating routine processes, organizations can redirect human talent toward creative problem-solving, strategic planning, and customer-centric initiatives.

RPA's relevance extends to the dynamic and interconnected nature of contemporary business environments. It seamlessly integrates with existing enterprise systems such as Enterprise Resource Planning (ERP), Customer Relationship Management (CRM), and Supply Chain Management (SCM), orchestrating a symphony of efficient, interconnected processes. This interconnectedness fosters a holistic approach to business operations, breaking down silos, enhancing cross-functional collaboration, and creating an end-to-end view of the organizations. As we journey into the landscape of understanding RPA, it becomes evident that beyond its technical details, RPA embodies a transformative force that is shaping the contours of the modern digital enterprise.

Within the broader context of enterprise systems, it is essential to focus on the key players that are shaping this transformative domain and to explore the significant entities in the RPA space, delving into their roles, contributions, and the impact they are having on the integration of RPA into contemporary business environments. Key players leading and innovating in the RPA technology market are at the forefront of the RPA ecosystem, wielding considerable influence on the trajectory of this technology. UiPath, Blue Prism, Automation Anywhere, and Microsoft have not only established a commanding presence in the market but also been instrumental in pushing the boundaries of innovation. UiPath's intuitive and user-friendly automation platform, Automation Anywhere's intelligent automation, Blue Prism's robust digital workforce solutions, and Microsoft RPA's seamless integration capabilities and adaptability within the Power Automate framework exemplify the different approaches that define the key RPA players. Understanding their strategies and offerings provides fundamental insight into the prevailing trends and benchmarks within the RPA landscape.

Complementing the stalwarts are other relevant players and specialized providers that are bringing dynamism and specialization to the RPA space. Companies such as Pega, Kryon, and EdgeVerve Systems are carving niches within specific RPA domains. Pega's focus on end-to-end automation, Kryon's innovative use of Artificial Intelligence (AI), and EdgeVerve's intelligent automation solutions exemplify the diversity of approaches taken by other players.

The collaborative dimension of the RPA landscape is vital to its evolution. RPA players often engage in strategic alliances, partnerships, and mergers to enhance their capabilities and expand their market reach. UiPath's partnerships with Microsoft and SAP, Automation Anywhere's collaboration with Google Cloud, Blue Prism's alliance with IBM, and Microsoft RPA's strategic integrations within the broader Microsoft ecosystem are illustrative examples of how key players in the RPA arena are forging synergies. Analyzing the influence of key players on industry dynamics is crucial to understanding the broader implications of their actions. These players not only set technological standards but also influence market trends, customer expectations, and the overall trajectory of RPA adoption within enterprise systems.

The RPA landscape is a complex ecosystem, and understanding its dynamics requires an examination of real-world applications and success stories. A few examples from the experience of the first author as a consultant are given below.

First example: a leading healthcare provider aimed to enhance operational efficiency and reduce processing times in its complex processes. These processes included entering patient data, scheduling appointments and exams, and creating patient reports, among others. To achieve this, it implemented UiPath to automate a set of these processes. This implementation resulted in streamlined processes, a significant reduction in processing times, and improved accuracy in these critical operations. As a result, healthcare professionals were able to focus more on patient care, resulting in increased productivity.

In the second example, a manufacturing company wanted to improve its back-office operations for production processes, with the main goal of lowering production costs and downtime. By using an RPA solution this organization was able to automate tasks for its inventory management, quality control, equipment monitoring, and production planning and maintenance processes. This resulted in fewer errors in back-office operations, less downtime, improved data quality, and lower operating costs.

Third, a telecommunications company sought to enhance process efficiency, targeting primarily its back-office functions spanning finance, human resources, commercial operations, customer service, and network management. This comprehensive initiative allowed the organization to improve internal operations and external service delivery, resulting in faster and better customer support.

As a fourth example, a construction company implemented an RPA solution to improve its shared service center operations, primarily focused on finance and human resources. By implementing this technology, the organization was able to speed up month-end closing, improve management reporting, and make tax handling faster.

These cases illustrate how RPA is delivering tangible value to organizations across different industries and provide insights into the versatility, adaptability, and transformative potential of RPA solutions.

1.2.2 RPA in Action: Operating at the User Interface

RPA works by imitating human actions at the user interface level. This involves a detailed, step-by-step analysis of task automation procedures. RPA interacts with the user interface in a way that mimics human behavior. This interaction is a multi-step process that includes identifying interface elements, managing data entry and extraction, executing actions, and integrating with the user experience. The ability of RPA to adapt to various application interfaces is a key part of this imitation process. The first step in the interaction process is for the RPA tool to identify the elements of the user interface. This could be anything from buttons and text fields to drop-down menus and checkboxes. The RPA tool uses various techniques to recognize these elements, such as image recognition, OCR (Optical Character Recognition), or anchor-based recognition. This allows the RPA tool to understand the layout and structure of the user interface, much like a human user would. Once the RPA tool has identified the interface elements, it can perform actions. These actions are typically pre-defined in the automation script and can include things like clicking a button, entering text into a field, or selecting an option from a drop-down menu. The RPA tool executes these actions in a way that emulates human interaction, ensuring that the underlying system perceives them as if a human user was performing them.

RPA tools are designed to seamlessly integrate into the user experience. They operate in the background, performing tasks without disrupting the user. This allows the system to operate continuously, with the RPA tool handling routine, rule-based tasks, freeing up human users to focus on more complex, value-added activities.

Integrating RPA into the complex structure of enterprise systems serves as a transformative catalyst that can lead to improved operational performance. RPA's integration with key enterprise systems is a nuanced process that aligns automated tasks with the complex frameworks of ERP, CRM, and other core business systems. Importantly, RPA operates within the existing infrastructure without causing significant changes to the core systems. It functions as a standard user, ensuring a non-disruptive and seamless integration into the digital ecosystem of the organization.

As an agile orchestrator, RPA adeptly manages data exchanges, executes tasks, and facilitates information flows within the established protocols of the core systems. The collaboration between RPA and existing technologies enhances the potential for improved performance. RPA does not operate in isolation; it integrates harmoniously with other technology frameworks. For example, combining RPA with AI enhances data analysis capabilities, providing organizations with deeper insights for informed strategic decision-making. In a fast-paced business landscape, characterized by rapid change and the importance of data, this synergistic alliance positions RPA as a strategic tool for organizations navigating the complexities of the modern business environment. It ensures that organizations can leverage the benefits of automation without requiring disruptive changes to their well-established operational structures.

The rise of RPA in industry adoption can be attributed to its inherent simplicity, speed, and cost-effectiveness. In their quest for operational excellence, organizations are drawn to the streamlined automation that RPA provides. Its user-friendly nature enables quick integration into existing processes, which is a key driver of its growing popularity among organizations.

Traditional methods often demand significant financial investments in lengthy development cycles, specialized coding expertise, and ongoing maintenance. In contrast, RPA, with its visual design interface and rule-based programming, not only reduces upfront costs but also minimizes the resources required for ongoing management. The agility afforded by RPA translates into accelerated task execution and improved responsiveness to dynamic business needs, making it a compelling choice for organizations seeking to improve efficiency and respond quickly to changing business dynamics without incurring additional costs.

RPA's cost-effectiveness extends across multiple dimensions. Its visual interface and pre-built modules significantly reduce development time, slashing the initial costs of deployment. Organizations can achieve rapid automation without exorbitant upfront expenditures. Moreover, the speed at which RPA executes tasks directly contributes to operational efficiency. Tasks that once required extensive manual labor are executed quickly and accurately, reducing labor costs and enhancing overall productivity. Additionally, as organizations grow, RPA scales seamlessly. Traditional methods may involve substantial costs to scale operations, whereas RPA allows for flexible scalability without proportional increases in expenditure. The adaptability of RPA to diverse business processes and systems also helps to reduce costs. It mitigates the need for extensive custom coding or system alterations, reducing associated costs and ensuring a faster return on investment.

1.3 ENHANCING BUSINESS PROCESSES: COST REDUCTION AND EFFICIENCY IMPROVEMENTS

In an era of rapid technological advancement and increasing reliance on digital processes, RPA is emerging as a valuable asset for organizations undergoing digital transformation.

The increasing complexity of business processes and the sheer volume of data handled by organizations necessitate a technology that can enhance efficiency without compromising accuracy. RPA's role in this context is not only about automation but also about augmenting human capabilities. It empowers organizations to ensure compliance and improve accuracy and efficiency.

1.3.1 Understanding the Impact of RPA on Business Processes

RPA plays a crucial role in automating complex tasks in organizations (Huang & Vasarhelyi, 2019). The role of RPA in organizations is akin to having a virtual workforce that can perform tasks with precision and speed, without the limitations of human constraints. It can handle tasks that are high volume and repetitive, such as data entry, invoice processing, and customer service queries. It not only expedites processes but also reduces the risk of errors associated with manual intervention. RPA's adaptability to diverse business processes makes it a versatile tool for enhancing operational efficiency and overall productivity within organizations.

RPA operates as a digital workforce, and understanding its mechanics is relevant to understanding its transformative impact on business processes. In elucidating the intricate workings of RPA, we embark on a detailed exploration, employing diagrams and flowcharts to render a visual narrative of the operational process (see Figure 1.1).

The diagram presented in Figure 1.1 is an example of the key components of an RPA process and serves as a fundamental guide. Importantly, it is adapted from a well-established and widely used framework used in UiPath. The importance of the

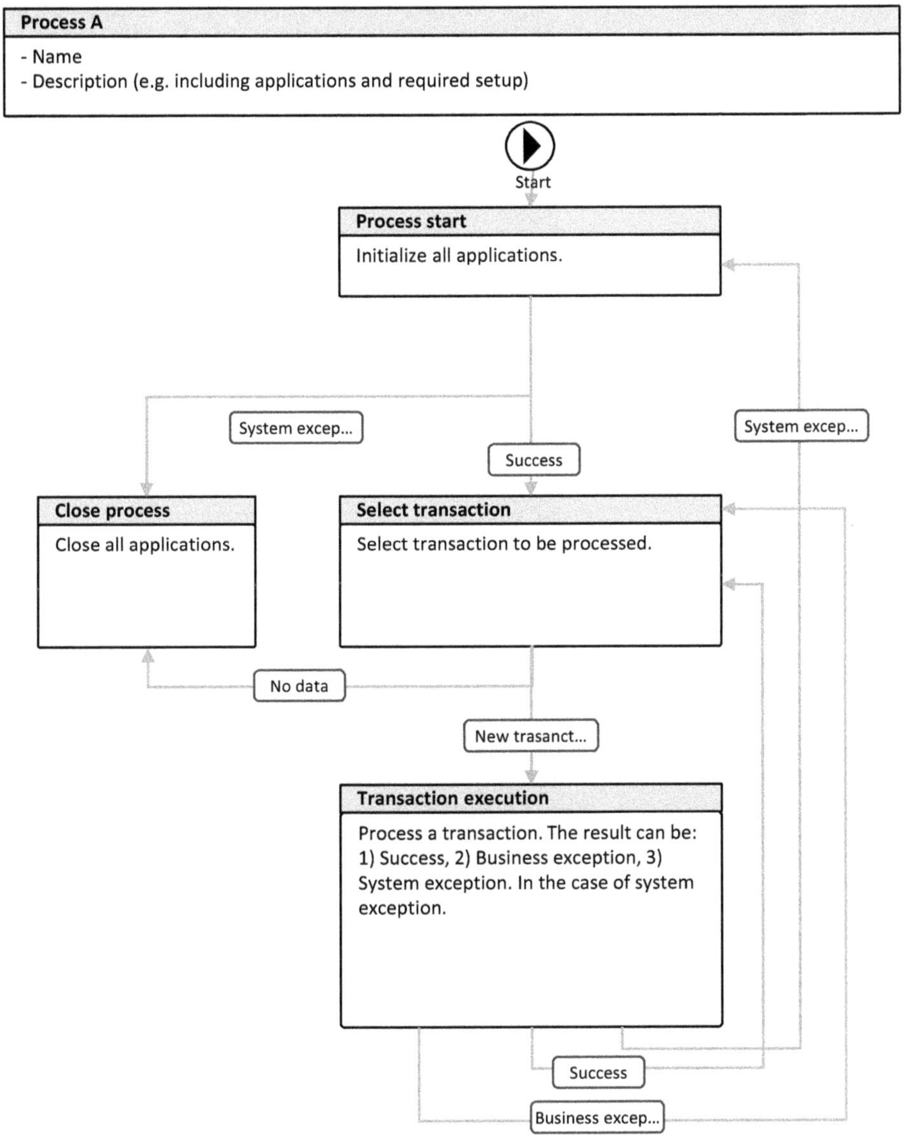

FIGURE 1.1 RPA framework diagram.

(Source: Adapted from UiPath.)

UiPath RPA Framework lies in its solution robustness, versatility, and user-friendly design, making it the preferred choice for many organizations embarking on their automation journey. By adhering to a well-conceived RPA framework, organizations can ensure a structured and effective approach to implementing RPA.

As mentioned above, RPA excels at many tasks. It seamlessly automates data entry tasks, ensuring accuracy and speed when processing large amounts of data. RPA navigates data input forms, validates information, and populates databases with precision. It also handles transactional processes, whether in financial transactions, order processing, or inventory management. It executes rule-based decisions, verifies transactions, and updates records, expediting overall processing times. Additionally, RPA efficiently manages records by automating document handling, data extraction, and archiving processes. It ensures the systematic organization and retrieval of information, reducing the risk of errors associated with manual record-keeping.

The visual aids provided by diagrams and flowcharts used during the automation cycle facilitate a comprehensive understanding of how RPA operates in the context of business processes.

The RPA framework can also be used to automate a group of tasks, such as the following example:

1. Start	Trigger: New customer registration.
2. User interface interaction	Identify UI elements: Collect customer details, including name, contact information, and account preferences.
3. Rule-based decision	Decision point: Check if the customer exists in the system. If yes, proceed to step 4. If no, initiate the account creation process.
4. Data exchange	Communicate with the CRM system. Input: Validated customer details. Output: Create customer profile.
5. Task execution	Send automated email notification to the customer welcoming them. Generate account credentials and send them to the customer.
6. Review and verification	Internal verification of customer data. Compliance check for regulatory requirements.
7. Approval process	Managerial approval for special account privileges. Auto-approval for standard accounts.
8. End	Trigger: Customer onboarding completed.

1.3.2 RPA's Contribution to Cost Reduction and Efficiency Improvements

In the context of business processes, cost reduction refers to strategies implemented by companies to reduce the cost of running their processes and to improve profitability. This can involve reducing overhead costs, renegotiating contracts, or implementing new technologies to automate manual tasks. Cost reduction involves minimizing the financial resources required to carry out business processes. This encompasses a strategic evaluation of operational expenses to optimize resource allocation without compromising the quality and effectiveness of processes.

Efficiency improvements, on the other hand, refer to the process of doing more with less – producing the highest quality output with the least amount of input. This can involve streamlining operations, eliminating waste, and improving productivity and performance. Efficiency improvements focus on enhancing the speed, accuracy, and overall effectiveness of business processes. This entails streamlining workflows, minimizing delays and errors to ensure that tasks are accomplished with the best use of resources.

Both cost reduction and efficiency improvements are critical for maintaining competitiveness and ensuring the long-term sustainability of a business.

RPA can contribute significantly to cost reduction in several ways (Deloitte, 2017). First, it reduces labor costs by automating repetitive, rule-based tasks that were previously performed by humans, allowing employees to focus on more strategic, higher-value tasks. Additionally, RPA minimizes operational overheads associated with manual data entry, processing, and record-keeping. By automating routine tasks, the need for extensive manual oversight is eliminated, leading to a reduction in operational costs. Moreover, RPA bots are programmed to follow rules and do not make human errors. Error mitigation helps to reduce costs by preventing expensive mistakes and the subsequent need for corrective action.

In terms of efficiency improvement, RPA executes tasks with remarkable speed, expediting processes that would traditionally be time-consuming. The execution of rule-driven tasks enhances overall process efficiency and reduces turnaround times. RPA also operates with a high level of accuracy, minimizing errors in data entry, processing, and decision-making. Improved accuracy not only enhances the quality of outcomes but also reduces the need for rework and corrections. RPA bots can work 24/7 without breaks, and they can complete tasks faster than humans, and their continuous availability helps improve efficiency by ensuring uninterrupted task execution and faster cycle times. RPA can improve compliance by maintaining detailed logs of all actions taken, reducing the risk of noncompliance and the potential costs associated with it. Finally, it can easily be scaled up or down to meet changing business needs, without incurring significant additional costs.

In summary, RPA positions itself as a strategic enabler for organizations seeking to enhance their operational capabilities while maintaining fiscal prudence. The transformative impact of RPA extends beyond mere automation; it fundamentally reshapes the cost efficiency dynamics of modern business operations.

Several examples are presented below, drawn from the experience of the first author as a consultant.

Customer Service: A company was handling approximately 10,000 calls per month, with each call taking approximately 8 minutes due to the complexity of a specific system. This led to inefficiency and a poor customer experience. After implementing RPA, the tasks were completed in less than 60 seconds, representing an 87% reduction in execution time.

Finance and Accounting: Several organizations have implemented RPA technology in their finance and accounting processes and have seen early success in their journey. RPA is helping to bridge the gap between different applications present in the legacy system, for example, automatically processing invoices or creating purchase orders.

Reconciling Transactions: A company needed to reconcile a high volume of payments for parking services daily. This repetitive task required a lot of staff time. After implementing RPA, a bot was able to access a master file of account information, download the bank movements of each account, apply a pre-defined set of business rules, and if there was a discrepancy, generate an Excel spreadsheet with a proposed reconciliation and send it to the finance team.

Employee Onboarding: Employee onboarding is typically a manual and time-consuming process. But with RPA and other automation tools, it can be much faster and easier. When an employee accepts an offer and sets a start date, a digital workflow using RPA can trigger the necessary steps to get them started.

In each of these examples, RPA has been used to automate different types of processes, resulting in increased overall productivity and efficiency, as well as cost savings and improved customer satisfaction.

1.3.3 Measuring the Impact of RPA in Business Processes

Many organizations lack sufficient understanding of the Key Performance Indicators (KPIs) associated with RPA programs that are crucial to the success of automation activities. By providing insights into RPA KPIs, this aims to support the enhancement of RPA programs and enable organizations to maximize the benefits of automation.

There are various metrics and KPIs that can be used to measure the impact of RPA. Here are some of the most common:

- Cost Savings: This is the reduction in costs as a result of implementing RPA. It can be calculated by comparing the costs before and after RPA implementation.
  - KPI 1: Labor Cost Reduction – Measure the percentage reduction in labor costs associated with tasks automated by RPA. Calculate the difference in the labor cost before and after RPA implementation, taking into account salaries, benefits, and associated overheads.
  - KPI 2: Operational Overhead Reduction – Assess the impact of RPA on overall operational overheads. This includes the reductions in expenses related to manual oversight, additional staff, and infrastructure costs.
- Time Savings: This measures the reduction in the time taken to complete tasks after implementing RPA. It can be calculated by comparing the time taken to complete tasks before and after RPA implementation.
  - KPI 3: Task Execution Time – Quantify the time taken to complete specific tasks before and after RPA implementation. Evaluate the reduction in task execution time, emphasizing the speed and efficiency gains achieved through automation.

    - KPI 4: Process Cycle Time – Analyze the total cycle time of key business processes. Measure the time it takes to start and finish a process and identify areas where RPA is speeding up the workflow.
- Error Rate Reduction: This measures the reduction in errors in tasks after implementing RPA. It can be calculated by comparing the error rates before and after RPA implementation.
    - KPI 5: Error Rate – Measure the error rate in tasks automated by RPA and compare it with the error rate in manual processes. A lower error rate indicates the effectiveness of RPA in ensuring accuracy and reliability.
    - KPI 6: Error Resolution Time – Assess the time taken to identify and rectify errors. Determine how swiftly RPA detects and addresses errors, contributing to a more responsive and streamlined error management process.
- Employee Productivity: This measures the increase in employee productivity following the implementation of RPA. It can be calculated by comparing the output of employees before and after RPA implementation.
    - KPI 7: Employee Workload – Evaluate the workload of employees involved in tasks automated by RPA. Assess the redistribution of responsibilities and gauge how RPA enables employees to focus on higher-value, strategic activities.
    - KPI 8: Employee Satisfaction – Implement surveys or feedback mechanisms to gauge employee satisfaction with the introduction of RPA. A positive shift in employee morale and job satisfaction can be an indicator of improved productivity.

Setting up a measurement system involves defining the metrics and KPIs, fixing targets, and establishing a process for collecting and analyzing data. An example of how this can be accomplished is as follows.

First, organizations should start by defining the metrics and KPIs that are relevant to their goals, such as cost savings, time savings, error rate reduction, and employee productivity. Once these metrics have been defined, realistic targets should be set for each metric to align with the organization's objectives. Next, a process for collecting data on each metric and KPI should be established. This could involve setting up automated data collection systems, conducting surveys, or using other data collection methods. Integration with RPA platforms can automate the data collection process, enhancing efficiency and accuracy. Once the data is collected, it can be analyzed to assess the impact of RPA on business operations. This involves comparing the data against set targets, identifying trends, and pinpointing areas for improvement. Iterative improvement is essential, as organizations should continuously refine the measurement system based on feedback and evolving business needs. Regularly reassessing the relevance of selected metrics and KPIs ensures that the measurement system remains effective.

The goal of measuring the impact of RPA goes beyond proving its value; it is also about identifying areas for improvement and maximizing the return on investment. Therefore, organizations should regularly review and update the measurement system to adapt to changes in the company and its needs. By adopting a comprehensive approach to metrics and KPIs, organizations can establish a systematic and data-driven framework for assessing the impact of RPA.

1.4 WORKFORCE IMPACT AND HUMAN–MACHINE COLLABORATION

RPA is transforming the workforce in unprecedented ways. As a technology that automates tasks once performed by humans, RPA is changing the nature of work by shifting the human workforce from manual, mundane tasks to more strategic and creative roles. This shift not only improves efficiency and productivity but also increases job satisfaction and employee engagement. The advent of RPA brings a paradigm shift that challenges traditional notions of work and task execution. Organizations that embrace RPA are experiencing a notable evolution in the nature of work, characterized by a strategic realignment of human roles.

RPA complements human skills in several ways. For instance, in a customer service department, RPA can handle tasks such as sorting emails, routing requests, and updating customer records. This allows customer service representatives to focus on understanding customer needs, solving complex problems, and building relationships with customers. Moreover, RPA can also handle tasks that require high accuracy and speed, such as data entry and transaction processing. This not only improves efficiency but also reduces the risk of human error, resulting in higher quality outcomes.

RPA can also facilitate collaboration in several ways. It can act as a bridge between different systems and teams, streamlining communication and coordination tasks. For example, in a supply chain management process, RPA can automate the exchange of information between the procurement, warehousing, and transportation teams. This ensures that all teams have access to up-to-date information, leading to better coordination and efficiency.

The implementation of RPA raises several ethical considerations. One of the main concerns is the potential for increased unemployment as some jobs become obsolete due to this new technology. Acknowledging these concerns is crucial for fostering a responsible and ethical approach to automation. Another ethical concern is privacy. RPA bots often handle sensitive data, and there are concerns about how it is collected, used, stored and protected. Organizations need to ensure that they have robust data protection measures in place and that they are transparent about how they do this.

Managing workforce transitions in the era of RPA is a complex task that requires careful planning and execution. Organizations are using different strategies to execute this process effectively. One strategy focuses on providing training and support to help displaced workers transition to new roles. This could involve reskilling initiatives, where workers are trained in new skills, or upskilling initiatives, where workers are trained to perform higher-level tasks. By investing in upskilling and reskilling programs, organizations empower employees to acquire the skills needed for new roles that meet the evolving needs of the workforce. This ensures a smoother transition and mitigates the impact of job displacement. Another strategy involves creating new roles and moving displaced workers into them. For instance, an employee who was previously involved in data entry could be moved to a role involving data analysis. An ethical approach involves recognizing the unique strengths of human workers and creating new roles that leverage these skills. Organizations can strategically design roles that capitalize on these distinctly human attributes and ensure the continued relevance of the workforce. One such approach is to design and implement a phased transition plan. Instead of implementing RPA all at once, organizations can opt for a gradual transition plan. This gives employees time to adjust to the new changes and reduces the shock of sudden job displacement. Rather than abrupt changes, organizations can phase in automation, allowing employees time to adapt and acquire new skills. Maintaining transparent communication throughout the transition process is also an essential strategy. Organizations should openly discuss the reasons for implementing RPA, the expected impact on the workforce, and the strategies in place to manage these changes. Transparency builds trust and addresses ethical concerns. Organizations can develop an ethical decision-making framework to guide RPA implementation. This framework should consider the social, economic, and cultural impacts of automation to ensure that decisions are consistent with ethical principles and prioritize the well-being of the workforce.

1.5 CHALLENGES AND FUTURE TRENDS OF RPA

While RPA offers numerous benefits such as increased efficiency, cost savings, and improved accuracy, its implementation is not without challenges such as resistance to change, lack of understanding of RPA, technical issues, dependence on existing infrastructure, and inability to handle non-routine tasks (Deloitte, 2017). These challenges need to be effectively addressed to ensure a successful RPA implementation.

Like any new technology, RPA can be met with resistance from employees who may fear job loss or who are accustomed to traditional workflows and may be concerned about disruption to their routine tasks. Organizations need to communicate transparently about the objectives of RPA implementation, emphasizing that automation aims to enhance efficiency rather than replace human roles. Real-world examples showcasing successful transitions and positive outcomes can serve as powerful motivators to reduce resistance.

There may be a lack of understanding of what RPA is and what it can do, leading to unrealistic expectations or misconceptions about the technology. Stakeholders, including decision-makers and end-users, may not fully comprehend the potential of RPA or how it aligns with organizational goals. This lack of awareness can impede buy-in and hinder the seamless integration of RPA into existing processes. Addressing this lack of understanding requires comprehensive education and awareness programs. Organizations should invest in training sessions that elucidate the capabilities of RPA, its impact on daily workflows, and the potential benefits to employees. Real-world case studies can be used to illustrate instances where RPA has augmented operational efficiency, resulting in positive outcomes for both the organization and its workforce. Technical issues can include integration difficulties with existing systems, lack of IT infrastructure to support RPA, issues with the RPA software itself, and software compatibility issues. The dynamic nature of the technology requires a careful evaluation of existing IT frameworks to ensure compatibility and efficient integration. Real-world examples can highlight instances where technical challenges have been overcome through strategic planning, collaboration with IT experts, and phased implementation. RPA bots interact with applications in the same way a human user would, through the user interface. This means they depend on the existing IT infrastructure and can be affected by changes to the applications they interact with. For instance, if an application is updated and its user interface changes, the RPA bot may no longer be able to interact with it effectively. Legacy systems or intricate IT frameworks can pose challenges to seamless integration. The reliance on existing infrastructure underscores the importance of pre-implementation assessments to identify potential bottlenecks and ensure a harmonious fusion of RPA with the organization's technological ecosystem. As RPA is best suited for automating repetitive, rule-based tasks, it's less effective for tasks that require human judgment, creativity, or decision-making. For instance, while RPA can handle tasks such as data entry or invoice processing, it cannot handle tasks like strategic planning or customer relationship management.

While the challenges and limitations of RPA are significant, they can be mitigated with the right strategies. These strategies involve careful planning, communication, and continuous learning and improvement.

Before implementing RPA, it's important to conduct a thorough feasibility study. This involves identifying the tasks that are suitable for automation, assessing the potential benefits and costs, and evaluating the technical and organizational readiness for RPA. Implementing RPA requires a clear strategy that includes defining the goals of automation, selecting the right tasks to automate, and planning the implementation process. A clear and well-defined implementation strategy is foundational to successful RPA adoption. Organizations need to define their objectives, identify suitable processes for automation, and establish a phased approach for integration. A strategic roadmap ensures that RPA aligns with organizational goals and minimizes disruptions during implementation. Employee involvement is crucial for the success of RPA implementation. This includes communicating the benefits of RPA to employees, addressing their concerns, and involving them in the planning and implementation process. Actively engaging employees in the process, seeking their input, and addressing their concerns can foster a collaborative atmosphere. By incorporating feedback from end-users, organizations can tailor RPA solutions to align with the unique requirements of their workforce, increasing adoption and reducing resistance. To ensure that employees can work effectively with RPA bots, it's important to provide adequate training and support. This could include training programs to equip employees with the necessary skills and a support system to help them troubleshoot issues and learn from their experiences. Ongoing support mechanisms, including help desks and troubleshooting resources, will help to maintain employee confidence and competence in working with RPA bots. Technical issues can be addressed through a robust IT support system. This involves having a dedicated IT team to troubleshoot issues, regular maintenance and updates of the RPA software, and continuous monitoring of the RPA processes. Technical issues can be proactively managed through robust troubleshooting mechanisms. Establishing dedicated IT support, engaging with RPA vendors for expertise, and fostering collaboration between IT professionals and end-users can expedite the resolution of technical challenges. Regular assessments and updates will also help keep RPA systems running smoothly.

The landscape of RPA is dynamic, with continuous advancements shaping its trajectory. Several emerging trends are shaping the future of RPA, such as the integration of RPA with AI and Machine Learning (ML), the development of more sophisticated RPA bots and cloud-based RPA solutions, and advances in the way RPA bots are trained and managed (Deloitte, 2015, 2017).

This integration of RPA with AI and ML is one of the most significant trends. It is leading to the development of intelligent automation solutions that can handle not only rule-based tasks but also tasks that require decision-making and learning from past data. This synergy empowers RPA systems with cognitive capabilities, allowing them to comprehend unstructured data, adapt to dynamic scenarios, and make decisions. Additionally, its integration with natural language processing is allowing to automate end-to-end business processes. RPA systems are evolving into intelligent agents that can learn, adapt, and execute complex tasks. The integration of natural language processing and advanced analytics is driving RPA to improve problem-solving and decision-making, transcending the boundaries of rule-based automation.

Another significant trend is the development of more sophisticated RPA bots. They are now capable of handling exceptions and making decisions based on predefined criteria. The sophistication of RPA bots is improving, driven by advancements in algorithms and computing power. Modern RPA bots have enhanced capabilities to navigate complex workflows, interact with disparate systems, and execute tasks with a greater degree of autonomy. The evolution of RPA bots positions them as versatile assets that can address a broader spectrum of organizational processes.

Furthermore, the development of cloud-based RPA solutions has made it easier for organizations to implement and scale RPA. These solutions offer benefits such as lower upfront costs, faster deployment times, and easier scalability. The advent of cloud-based RPA solutions marks a paradigm shift in deployment models. Cloud-based RPA offers scalability, flexibility, and accessibility, allowing organizations to deploy automation solutions without the constraints of on-premises infrastructure. This evolution aligns with the broader trend of cloud adoption, facilitating seamless integration and management of RPA processes.

Advances in the way RPA bots are trained and managed are significant. These include the use of AI and ML to train bots and the development of sophisticated bot management platforms that make it easy to monitor, manage, and scale RPA bots. Enhanced management platforms provide organizations with comprehensive insights into bot performance, enabling proactive adjustments, troubleshooting, and optimization of automation workflows.

To sum up, the field of RPA is rapidly evolving, with new trends and advancements emerging all the time. These developments are expanding the capabilities of RPA and opening up new opportunities.

1.6 CONCLUDING REMARKS

RPA is not just a technological tool; it is a strategic enabler for organizations seeking a competitive edge. It is a transformative technology that is changing the way organizations approach routine, rule-driven tasks. RPA uses software "robots" to automate traditionally human-led processes by interfacing with existing computer systems. Its widespread adoption is due to its simplicity, speed, and cost-effectiveness compared to traditional methods. RPA integrates seamlessly with existing enterprise systems such as ERP, CRM, and SCM, making the processes more connected and efficient. This interconnectedness fosters a holistic approach to business operations, breaking down silos, improving cross-functional collaboration, and creating an end-to-end view of the organizations. RPA operates within the existing infrastructure without causing significant changes to the core systems, functioning as a standard user and ensuring a non-disruptive and seamless integration into the digital ecosystem of the organization.

It streamlines complex tasks, reduces operational costs, and enhances accuracy by efficiently handling data entry, report generation, and other tasks. However, its implementation is not without challenges, including technical issues, organizational resistance, and the need for ongoing maintenance and updates. The future of RPA is rapidly evolving, with new trends and advancements continually emerging to expand its capabilities and open up new possibilities for its application.

REFERENCES

Ahlstrom, D., Arregle, J. L., Hitt, M. A., Qian, G., Ma, X., & Faems, D. (2020). Managing technological, sociopolitical, and institutional change in the new normal. *Journal of Management Studies*, *57*(3), 411–437.

Bundy, J., Pfarrer, M. D., Short, C. E., & Coombs, W. T. (2017). Crises and crisis management: Integration, interpretation, and research development. *Journal of Management*, *43*(6), 1661–1692.

Deloitte (2015). Automate this. The business leader's guide to robotic and intelligent automation. https://www2.deloitte.com/content/dam/Deloitte/ca/Documents/consulting/ca-en-the-robots-are-here-aoda.pdf

Deloitte (2017). The robots are ready. Are you? Untapped advantage in your digital workforce. https://www2.deloitte.com/pt/en/pages/strategy-operations/articles/robotic-process-automation.html

Deloitte (2023). Supporting digital transformation with RPA. Rethinking processes for seamless operations. https://www2.deloitte.com/mt/en/pages/rpa-and-ai/articles/digital-transformation-and-robotic-process-automation.html

Donthu, N., & Gustafsson, A. (2020). Effects of COVID-19 on business and research. *Journal of Business Research*, *117*, 284–289.

Enriquez, J. G., Jimenez-Ramirez, A., Dominguez-Mayo, F. J., & Garcia-Garcia, J. A. (2020). Robotic process automation: A scientific and industrial systematic mapping study. *IEEE Access*, *8*(1), 39113–39129. https://doi.org/10.1109/ACCESS.2020.2974934

Hällgren, M., Rouleau, L., & De Rond, M. (2018). A matter of life or death: How extreme context research matters for management and organization studies. *Academy of Management Annals*, *12*(1), 111–153.

Huang, F., & Vasarhelyi, M. A. (2019). Applying robotic process automation (RPA) in auditing: A framework. *International Journal of Accounting Information Systems*, *35*, 100433.

Koch, T., & Windsperger, J. (2017). Seeing through the network: Competitive advantage in the digital economy. *Journal of Organization Design*, *6*, 1–30.

Sharma, P., Leung, T. Y., Kingshott, R. P., Davcik, N. S., & Cardinali, S. (2020). Managing uncertainty during a global pandemic: An international business perspective. *Journal of Business Research*, *116*, 188–192.

Sharma, T. K. (2023). Hybrid Working: The Future of Organizations. In *Reshaping the Business World Post-COVID-19* (pp. 41–68). Apple Academic Press.

Soto-Acosta, P. (2024). Navigating uncertainty: Post-pandemic issues on digital transformation. *Information Systems Management*, *41*(1), 20–26.

Thakur, R., & Hale, D. (2022). Strategic crisis response: Managerial implications and direction for recovery and survival. *Journal of Business & Industrial Marketing*, *37*(10), 1959–1973.

Winston, A. (2020). Is the COVID-19 outbreak a black swan or the new normal? *MIT Sloan Management Review*, *16*. https://sloanreview.mit.edu/article/is-the-covid19-outbreak-a-black-swan-or-the-new-normal

2 Mobilising ERP and BI for Management Control

Managing Tensions and Hybridity

Bárbara Corvacho and João Oliveira
University of Porto, Porto, Portugal

2.1 INTRODUCTION

Corticeira Amorim (CA), the cork industry's world leader, launched a transformational, technological and functional project in 2016 to modernise and standardise processes and implement a new Enterprise Resource Planning (ERP), common to the group's various Business Units (BUs). Around the same time, CA also implemented a new Business Intelligence (BI) tool. Upon completion of the two projects, which structurally changed the group's information systems (IS), this study analyses the implementation of these projects and their repercussions on management control (MC), given this organisational function's reliance on IS to fulfil its aim of collecting, analysing and sharing information and assisting in the decision-making process (Rikhardsson & Yigitbasioglu, 2018).

ERP and BI implementations continue to expand (Ain et al., 2019; Katuu, 2020), and the associated literature, in particular involving the MC function, is heterogeneous in terms of the focus of analysis and perspectives, with little consensus (Dechow & Mouritsen, 2005; Spraakman et al., 2018). Implementation is acknowledged to be highly complex, and although many studies have already focused on implementation issues, their diverging results suggest that the organisational repercussions may be dependent on specific contextual and situated factors (Dechow & Mouritsen, 2005; Goretzki & Messner, 2019; Rautiainen et al., 2024; Spraakman et al., 2018). The infrastructural, long-term (Dambrin & Grall, 2023) and integrative nature of these systems signals the relevance of simultaneously studying both (Fähndrich, 2023), how their implementation is carried out and followed up, by whom, with what objectives and under which knowledge templates (Berlinski & Morales, 2024; Lodh & Gaffikin, 2003), with particular attention on MC's participation and changes in MC's tasks and

 DOI: 10.1201/9781003461821-2

roles (Byrne & Pierce, 2007; Fähndrich, 2023; Rautiainen et al., 2024; Sánchez-Rodríguez & Spraakman, 2012; Wolf et al., 2020). Thus, two intertwined research questions were defined: "Why and how were an ERP and a BI system implemented at Corticeira Amorim?" and "Why and how has management control changed at Corticeira Amorim?"

To answer these questions, this study adopts a qualitative and interpretive approach and the sensitising lenses of the "people, processes and technology" framework. The study explores the motivations for the change, the implementation process, and the repercussions in the MC area, particularly its tasks, roles and required skills. Motivations and repercussions of changes are analysed in an intertwined way, since MC both affects and is affected by the technological changes, within (and affecting) the particular organisational context. The study analyses the impacts, benefits and drawbacks focusing on the implementation project team, the MC department and the Executive Management; in particular, it analyses how the confluence of distinct objectives, perspectives and knowledge held by different actors (controllers and technologists) played out in the systems' implementation and operation. In this way, the study covers the multiple organisational areas involved, in a selected set of companies from this group.

Section 2.2 reviews the literature, Section 2.3 describes the methodology and the companies studied and Section 2.4 analyses the empirical insights. Section 2.5 deepens the discussion by exploring the consequences upon intra-organisational relationships and dependencies, and the confluence of different knowledge templates. Section 2.6 presents the conclusions, acknowledges limitations and suggests research avenues.

2.2 LITERATURE REVIEW

2.2.1 ERPs, BIs and Management Control

Information systems are essential for MC to provide information. These systems consist of a set of information technology (IT) infrastructures, application systems and employees who use IT to produce structured information (Baskerville et al., 2020). ERPs are promising to MC, given their aim to integrate companies' functions, processes and information flows (Fähndrich, 2023; Nofal & Yusof, 2013; Sánchez-Rodríguez & Spraakman, 2012) in a single system and a centralised database, shared with and accessible to authorised employees (van Roekel & van der Steen, 2019). An ERP can also integrate processes from several companies into one single ERP (Chou et al., 2005), providing group-level members access to all the group information. However, while ERPs excel in integrating transactional data, they tend to lack in-depth analytical and reporting capabilities (Chou et al., 2005; Rikhardsson & Yigitbasioglu, 2018).

The more recent adoption of BI systems has provided additional capabilities to aggregate data, from within a company, an entire group and even beyond organisational borders, and analytical capabilities to flexibly explore that data (Ain et al. 2019; Appelbaum et al., 2017; Chen et al., 2012; Oesterreich et al., 2019). BIs improve information timeliness, reliability and transparency, offering enhanced, multidimensional reporting and decision-making support and facilitating financial audits (Oliveira & Ribeiro, 2023; Power & Sharda, 2015; Rikhardsson & Yigitbasioglu, 2018; Yeoh & Koronios, 2010). These BI capabilities make them ideal for addressing the above limitations of ERPs, so combining ERPs' data management and BIs' analytical processing is often recommended (Fähndrich, 2023; Rikhardsson & Yigitbasioglu, 2018; Sánchez-Rodríguez & Spraakman, 2012). From a research point of view, this emphasises the relevance of including both systems in technological impact studies on controllers' roles (Chou et al., 2005; Oliveira & Ribeiro, 2023), as done in this study.

Potential benefits for MC derive from the numerous functions of these systems, the centralisation, flow and transparency of information, even across international locations (Babaei et al., 2015; Sánchez-Rodríguez & Spraakman, 2012; Sheu et al., 2004), more relevant, timely and reliable information and better support for analytical and decision-making processes (Ain et al., 2019; Appelbaum et al., 2017; Granlund & Malmi, 2002; Oesterreich et al., 2019; Sánchez-Rodríguez & Spraakman, 2012; van Roekel & van der Steen, 2019; Wolf et al., 2020). They can function as an MC tool for operational, financial and strategic control, through standardised systems in factories and subsidiaries (Teittinen et al., 2013), adapted to the needs of each employee. MC and IS such as ERPs and BIs can even be considered as inseparable entities (Teittinen et al., 2013; Berlinski & Morales, 2024).

These capacities of ERPs and BIs have fuelled the potential of changes in the role of the controller, progressing from a "bean counter" absorbed in transactional recording processes to a "business analyst" or "business partner", increasingly involved in analysing information and participating in strategic and operational decision-making processes (Goretzki & Messner, 2019; Oliveira & Ribeiro, 2023; Rautiainen et al., 2024; Sánchez-Rodríguez & Spraakman, 2012).

However, the actual impact on MC may be only moderate or lag in time. This may be explained by factors such as changes in roles and identities typically happening only gradually (Fachin & Langley, 2024; Granlund & Malmi, 2002; Kallunki et al., 2011; Rautiainen et al., 2024), by the long periods during which these technological objects remain and evolve in organisations (Dambrin & Grall, 2023), and by the lack of technological determinism (Berlinski & Morales, 2024; Fähndrich, 2023; Rautiainen et al., 2024; Wolf et al., 2020). For example, Oliveira and Clegg (2015) showed how changes in management and organisational control only emerged from the combined and reinforcing effects of a technological innovation (an ERP) and two organisational innovations (e.g., a Shared Service Centre and a Corporate Centre).

IS-driven changes in controllers' roles vary and depend on multiple contextual and situated factors (Dechow & Mouritsen, 2005; Fähndrich, 2023; Goretzki & Messner, 2019; Rautiainen et al., 2024). Controllers need to further develop knowledge and competencies, such as good communication and teamwork skills to promote collaboration, and knowledge of the business, management and accounting processes and new IS and technologies (Oliveira & Ribeiro, 2023; Rautiainen et al., 2024), an evolution found to be challenging (Appelbaum et al., 2017; Oesterreich et al., 2019). Paradoxically, ERPs and BIs can even lead to an increased closure on the "bean counter" role since richer, timely information becomes available not only to

MC but across the organisation without requiring controllers' episodic, direct interventions (Oliveira & Ribeiro, 2023). This could be worrying for MC by potentially reducing its mechanisms and opportunities to influence decisions and guide behaviour and overall losing control (Berlinski & Morales, 2024; Kallunki et al., 2011; Rautiainen et al., 2024).

This conquest of new roles requires that controllers take action and intervene. These interventions may occur in everyday practices or in critical "trials of truth", affirming controllers' identity (Lambert & Pezet, 2011). Fachin and Langley (2024) highlighted how identity work emerges and unfolds during interpersonal interactions, within situated interactions. However, these interventions can also occur at these systems' implementation stage since they will constitute structures (actually, infrastructures – Yeoh & Koronios, 2010) that enable or constrain practices and possibilities of evolution (Dechow & Mouritsen, 2005). Therefore, controllers may attempt to structurally embed suitable characteristics in these systems to enable the future accommodation of their perspectives and objectives (Berlinski & Morales, 2024). The ensuing section will, therefore, focus on implementation.

2.2.2 Implementing ERPs and BIs

Implementing an ERP is more than just a change of a technological solution since it is both a commitment to change and a journey of organisational development (Teittinen et al., 2013), drastically transforming the company and envisaging making it more efficient and effective (Rautiainen et al., 2024). Implementation can be done in a Big Bang approach or a phased manner, one module at a time (Katuu, 2020). However, it is typically a long and arduous process, and not all companies are able to reap the benefits.

Critical success factors have long been identified. Ehie and Madsen (2005) identified the correct definition of the strategy by top management, including support and continuous monitoring, and also project management, including the correct definition of the project scope and the establishment of a project team with well-defined objectives and tasks. Bradley (2008) highlighted the importance of employee training, the project manager and the power user (an employee who understands both the technological and the business aspects and inspires other employees), and Grainger and McKay (2007) identified the company's culture as a critical factor. On the other hand, failure can result from a lack of support from top management, employee resistance, poor understanding of business requirements and implications, lack of maturity about integration or conflicting objectives between IT and business (Momoh et al., 2010; Yusuf et al., 2004).

The literature emphasises the need for significant financial and human resources for ERP and BI system implementations, highlighting the need for careful data migration and employee training on accurate transactional recording to assure data quality and subsequent reliable reporting. However, data validation may be time-consuming (Berlinski & Morales, 2024), and there may be skill shortages or untraceable errors (Teittinen et al., 2013). Management issues, especially in aligning ERP with business operations and managing change, are deemed more critical than technical problems (Babaei et al., 2015; Sheu et al., 2004). Moreover, international ERP implementations face added hurdles due to linguistic, cultural, regulatory and skill diversity (Babaei et al., 2015; Sheu et al., 2004). For BI systems, challenges extend to integration with existing systems, data quality, security and privacy, emphasising the necessity of user engagement and a comprehensive approach that spans technical, legal and organisational dimensions (Chen et al., 2012; Yeoh & Koronios, 2010).

The above review of critical success (and failure) factors highlights the multitude of factors and actors involved in implementing a complex IS, such as an accounting IS, an ERP or a BI. Actors may be internal to the organisation, from areas such as business, IT or accounting/controlling, or external to the organisation, such as implementation and software vendor consultants (Lodh & Gaffikin, 2003). Berlinski and Morales (2024) argued that implementing an accounting IS implies the confluence between the business world (in particular, accountants and controllers) and the IT world (the technologists as systems experts), creating a blend of objectives, perspectives, and knowledge. This integration shapes the implementation process and outcomes, potentially leading to tensions between the business and IT goals. Berlinski and Morales (2024) emphasised the importance of understanding this dynamic interplay as another element of contextual contingency on the impact of ERP and BI upon the controllers' role, through deeper research of local and situated practices and greater emphasis on the often-neglected technologists' viewpoint.

2.3 METHODOLOGY

2.3.1 Research Questions, Framework and Method

The reviewed literature highlights the relevance and intertwinement of the two broad research questions. The first research question, "Why and how were an ERP and a BI system implemented at Corticeira Amorim?", characterises the context and motivation for the systems' implementation ("why?") and the implementation process itself ("how?"). This is a process widely acknowledged as complex, involving distinct actors' objectives, perspectives and knowledge templates and likely triggering tensions whose developments and outcomes require further research. Therefore, the first research question builds on the reviewed literature, culminating in Berlinski and Morales' (2024) focus on knowledge templates and tensions.

The second research question, "Why and how has management control changed at Corticeira Amorim?", builds on the reviewed literature on the changes in MC and controllers' tasks, roles and skills. This question is related to the first research question, since the literature has documented how IS and MC change processes may be imbricated and reciprocally influence each other, in non-deterministic processes involving multiple actors in various timeframes, both during and after implementation (Berlinski & Morales, 2024; Dechow & Mouritsen, 2005; Fähndrich, 2023; Goretzki & Messner, 2019; Oliveira & Clegg, 2015; Rautiainen et al., 2024; Wolf et al., 2020). Therefore, changes in MC can only be understood in their wider organisational context – a context which may both influence and be influenced by such changes.

This study also mobilised the "people, processes, and technology" framework, originated by Leavitt (1965), as sensitising lenses to analyse empirical data. This seminal and still widely used framework (Prodan et al., 2015) emphasises the importance of people in IS, highlighting their role alongside technology in process design. It offers a comprehensive view for implementing IS like ERPs and BIs by considering people, processes and technology together, hence suiting their intertwined nature in the context of accounting and control (Berlinski & Morales, 2024).

The defined research questions recommend adopting qualitative methods (Yin, 2018), as they are suitable for addressing "why" and "how" questions and enable a comprehensive and in-depth view of social phenomena (Yin, 2015). This exploratory and explanatory single-case study is based on the deliberate selection (Yin, 2015) of the case of CA, since in 2016 CA had begun a transformational process of transition and integration of ERPs into a single ERP, adopting the SAP solution. Until then, three different ERPs were used in CA, across and even within BUs, and there were companies with ERPs from the same supplier that worked independently. The implementation process was phased, and the five BUs already had at least one company using SAP at the time of the study (2021), even though the maturity levels in using the system were different. Furthermore, around the same time, CA started implementing a BI system.

This study investigates the motivations for this project, the implementation process and the changes in the three components of the "people, processes and technology" framework. It evaluates the project's benefits, challenges and overall success, examining the perspectives and roles of various actors, namely managers, controllers and the often-neglected group of "technologists" (Berlinski & Morales, 2024). Finally, the study focuses on the impact of these changes on MC and its tasks, roles and skills, as well as the intra-organisational relationships involving the MC department.

2.3.2 Information Generation Techniques

This research adopts an interpretive approach, using techniques such as structured interviews, document analysis and direct observation (Yin, 2018). The interview questions cover key issues in the "people, processes and technology" framework: in terms of people, the changes to their tasks, roles and skills; in terms of processes, the impacts of standardisation, systematisation and the new workflow of tasks; and in terms of technology, the infrastructure and systems in operation, the databases in use and preferred, and the main MC tools.

Three groups of interviewees were defined, based on their role and their relevance to the topic, to combine views associated with the project or company (e.g., its culture, implementation objectives and maturity of IS use) and views associated with the performed role (e.g., its use of the system or responsibilities). Group 1 (EM – Executive Management of BUs) provided a comprehensive overview of the project's motivations and impacts on MC. Group 2 (MC – MC of BUs) provided a more operational view of changes in the day-to-day work of controllers. Group 3 (PT – Project Team implementing the ERP and BI systems) provided a perspective on the projects, their motivation, challenges and benefits. This third group is a hybrid entity, combining technological and business characteristics, and enables following Berlinski and Morales' (2024) confrontation between technologists and BU MCs, while innovating by its hybrid form. The 11 interviews lasted between 40 and 70 minutes, and the

12 interviewees are identified by the abbreviations given above (PT, MC and EM), followed by a number (Table 2.1).

Multiple documents from project management and direct observation were used, as the first author worked for two and a half years as a functional consultant (i.e., a member of the PT) on the ERP project. To minimise bias, this text is based almost exclusively on empirics gathered explicitly for this purpose, that is, through interviews. However, the author's lengthy immersion in the organisation and projects

TABLE 2.1
Interviewees' Description

Interviewee Code	Interviewee Group	Company	Function	Role in the Project	Interview Date
EM1	Executive Management	Amorim Florestal Portugal	Finance Director	Project Director	12.07.2021
EM2	Executive Management	Amorim Cork	Head of MC	Process Owner	22.07.2021
EM3	Executive Management	Amorim Cork	Director of MC and Information Systems	Project Director	13.07.2021
EM4	Executive Management	Amorim Cork	CO-CEO	Project Steering	27.07.2021
MC1	Management Control	Amorim Florestal Portugal	Controller	End User	12.07.2021
MC2	Management Control	Amorim Florestal Portugal	Head of MC	Process Owner	14.07.2021
MC3	Management Control	Amorim Cork	Controller	End User	19.07.2021
MC4	Management Control	Amorim Cork	Head of Planning, Consolidation and Performance Management Systems	End User	14.07.2021
MC5	Management Control	Amorim Cork America	Finance Director	End User	13.07.2021
MC6	Management Control	Amorim Cork America	Chief Financial Officer	Process Owner	13.07.2021
PT1	Project Team (OASIS and ABBA)	Project	Functional Consultant	Functional Consultant	19.07.2021
PT2	Project Team (OASIS and ABBA)	Project	Team Leader and Global Process Owner	Team Leader and Global Process Owner	15.07.2021

enabled a deep, contextualised understanding of the empirics, particularly the organisation's people, processes and technologies.

The interviews were recorded and fully transcribed. The relevant insights were identified, related and annotated in an interactive way since each interview often contributed to developing the contextualisation and interpretation of insights from other interviews. More generally, empirical triangulation compared insights obtained from the various sources (interviews, documents and observation). This triangulation tended to reveal agreement and complementarity with regard to change processes, while diverging insights typically highlighted different interviewees' subjective standpoints and perceptions. These divergences, far from being considered problematic, reflected the challenges and tensions identified in the literature and enabled us to deepen our contextual understanding.

2.3.3 Selection of Empirical Objects

Two collective empirical objects were conceptualised for this research: the group BUs in which the systems were implemented and the PT involved in their implementation.

CA comprises five BUs, operating in a circular economy. Three of these BUs use raw materials that are co-products and by-products from upstream production processes. The remaining two are intrinsically linked by solid supplier–buyer relationships. Three companies were deliberately (Yin, 2018) chosen (Amorim Florestal Portugal – AFL; Amorim Cork – AC and Amorim Cork America – ACAM), for multiple reasons. First, they are interdependent in the value chain: AFL prepares the cork that AC will use, in Portugal, to produce natural cork stoppers and that, subsequently, ACAM will label, treat, package and distribute to the end customer throughout North America. Second, these companies had different systems; even though AFL and AC actually had an ERP from the same vendor, the information was not integrated, so in practice they had different systems. Finally, these companies had different Executive Management and MC departments, their new ERP and BI implementations were spread out over time and their stage of system usage maturity was different.

The two systems were implemented almost simultaneously: the ERP, through the project OASIS (One Amorim System and Integrated Solution), to uniformise and implement global processes through the implementation of SAP; and the BI, through the project ABBA (Amorim Better Business Analytics), to build analytical data models, using SAP as the primary source of information. The two projects had the same central team members, so both will be referred to as a single Project Team.

The project OASIS was divided into five specific functional areas (Costing & Analytics; Financial Accounting & Purchasing; Sales & Distribution; Industrial Data; and Master Data) and was constituted by a group of employees and external consultants entirely devoted to implementation. Each area had a team leader, with previous functional experience in the group, and hired functional consultants, who acquired extensive knowledge on the group business and processes through training and during the process design stage. As such, the PT was a hybrid entity, combining technological and business knowledge and perspectives, including the external consultants (with in-depth knowledge about the tools). Although the PT was the critical player in these projects, it had no hierarchical responsibility over the other actors;

hence outcomes emerged from interactions, rather than dictated through hierarchical power (Berlinski & Morales, 2024).

By contrast, ABBA was performed by one single team, assembling members from CA and an implementation partner. CA provided both functional members (from OASIS's Costing & Analytics team, given the strong connection of topics) and technical members. Additionally, the team also identified a responsible member for each BU, who would liaise between the business and the PT. Other stakeholders were involved if needed.

2.4 EMPIRICAL RESULTS

This chapter presents the empirical findings, structured according to the views of the two empirical objects defined above (the PT and BUs), concerning the motivations and overall results of the two projects, the change management processes and the controllers' tasks, roles and skills.

2.4.1 The Project Team's View

2.4.1.1 Motivations and Overall Results

PT interviewees agreed that the PT's main objective was to implement uniform global processes through a single ERP and a single BI, ultimately resulting in a global analytical model. Adopting the "people, processes and technology" framework, the following motivations for ERP integration could be identified: (1) difficulty in transferring skills internally; (2) failure, as a group, to have standardised processes and information and (3) obsolescence and diversity of the existing technological structure. "*As the processes were different and the information did not always represent the same thing, transferring people to different BUs was very difficult. The learning curve was very long and almost the same as recruiting from outside*" [PT2]. The interviewees also agreed that the objectives had been achieved. The technological structure was modernised entirely, with only some satellite applications still maintained, and the standardisation of processes led to clear synergies and standardisation of systems' skills (Sánchez-Rodríguez & Spraakman, 2012).

The BI project arose due to technological and functional factors. Interviewees highlighted the difficulty of adapting the technological architecture to the new data source (the ERP) and the possible inability of the existing architecture to support the information load: "*The SAP data model was disruptive (...) and the old BI might not have been able to deal with all the information, in volume and quality*" [PT2]. In functional terms, the new BI leverages upon the standardisation and reliability of information in the ERP by "*structured and standardised processes (...). And that is why we should also have the same BI and global [analysis] model*" [PT2]. As such, BI created an excellent capacity for structured exploitation of the new data generated by SAP, identical across BUs (Sánchez-Rodríguez & Spraakman, 2012).

SAP and BI systems were described as complementary but with different characteristics in terms of timeliness, granularity and the ability to aggregate and relate information, as well as in terms of their users (Nofal & Yusof, 2013; Rikhardsson & Yigitbasioglu, 2018; Sánchez-Rodríguez & Spraakman, 2012). "*Analysing the three organisational levels: the strategic level does not use SAP; the management level uses both, especially BI; and the operational level only works with SAP*" [PT2].

2.4.1.2 Change Management

CA's culture is widely considered open to change (cf. Grainger & McKay, 2007). Only in middle management did some resistance emerge, not associated with the technology but with the overall process design, due to the lack of a clear Global Process Owner (GPO) role or the overlapping positions of GPO and Team Leader.

> CA is very open to innovation and creativity (…) The only tricky part was deciding on the final solution (…) It could not be the sum of its parts, and it was not possible to satisfy all the BUs. This role should be that of a GPO, but that was not entirely clear, so the Team Leaders had to play that role.
>
> [PT2]

The organisational challenges were directly related to the willingness and ability to align, especially in the phase of designing global processes and aligning internal teams. These challenges had already been expected, but it was not expected that five years later they still would not be fully overcome.

Indeed, the complexity of the SAP database made it difficult to develop the solution. No negative consequences associated with the solution as a whole were identified, but some obstacles emerged, attributed mainly to the still low maturity of the system's use, namely some loss of application skills, the difficulty in understanding complex processes and the organisation's continuous demand for information. "*Even if the MC Department has not been involved in every process design, it is the MC that has to present and justify the results to the board of directors*" [PT2]. In MC, resistance to change was mitigated by previous projects to standardise the analytical model in CA, which outlined the way forward and facilitated the discussion and development of new processes during SAP design. As for the organisation as a whole, the alignment of top management was fundamental (Ehie & Madsen, 2005). "*Programmes like this can only be done with sponsorship from the top and from someone who conveys the capacity for change and an open mind*" [PT2].

The ABBA project was successful, but managing expectations was an issue. Some companies already had complete BI solutions, but they needed to recreate them from scratch again and hence recede in their maturity of BI use ("*We went back from a stage of consulting reports on mobile phones to designing models*" [PT2]).

Overall, the PT's interviewees rated the implementation as very successful, attributable to factors such as "*people's capacity for dedication, (…) project sponsorship, project management, clear alignment of the organisation and the project team, review and reassessment of objectives, change management, training*" [PT1], the natural openness to change, team spirit and involvement. In other words, the human and relational component was considered decisive for successful implementation (Ehie & Madsen, 2005; Yusuf et al., 2004).

2.4.1.3 Controllers' Tasks and Roles

PT's interviewees described a clear and positive change in MC's tasks. The standardised processes, integrated and systematised in the ERP, have promoted a more analytical role for controllers, to the detriment of an operational and transactional role (e.g., eliminating some tasks related to production processes) (Kallunki et al., 2011; Oliveira & Ribeiro, 2023; Sánchez-Rodríguez & Spraakman, 2012). "*MC has changed to a more analytical, interpretive area, rather than an operational one. And the role of business partner still needs to be further developed and extended, among the BUs and within MC itself*" [PT2] (Oesterreich et al., 2019). There was a clear separation between the finance department and the MC department, "*even in the division of responsibilities, in system entries and the ledgers themselves*" [PT2].

The analytical processes were improved and became entirely based on SAP, with no tasks outside the system. Profitability analysis became a consequence of the underlying transactional process, adopting a new process with previously impossible dimensions of analysis (Momoh et al., 2010): "*We used to measure performance inside our organisation a lot, but we did a very weak analysis of the outside – markets, countries, sectors, distribution channels and customers*" [PT2]. Finally, the only process associated with MC that still required a transactional record – period-end closing – was based on a task plan, the Schedule Manager, which systematises and orders all the steps to be carried out.

The implementation of SAP promoted the application of best practices. Processes were designed based on global standards, and the integration and complete flow of existing transactional processes in the ERP were ensured (Nofal & Yusof, 2013; Sánchez-Rodríguez & Spraakman, 2012; Teittinen et al., 2013). As a result, MC started obtaining the necessary, high-quality information to aid decision-making, without, for example, the need for month-end records to calculate the actual cost and subsequently estimate the value of stocks. However, and although the information in SAP became reliable and timely, the speed of the benefits was not as expected or desirable, especially in MC. As an area that "consumes" information produced by operational processes, it has to wait for these upstream processes to mature and to improve its ability to analyse this information (Appelbaum et al., 2017).

2.4.2 The Business Units' View

2.4.2.1 Motivations and Overall Results

The BUs' interviewees identified multiple motivations for the projects, related to limitations in the previous ERPs, such as slowness, lack of licences, interface problems, access difficulties, information gaps and the lack of a consolidated vision between departments (Baskerville et al., 2020): "*The worst thing was that each department only had its own vision, there was no consolidated vision, we all had different information*" [MC4]. This problem mainly affected AC, as a holding company, which managed several companies with different technological processes and infrastructures (Oliveira & Clegg, 2015). The technological solution in Portuguese companies did not meet organisational and information needs. "*CA was growing, but that was not the case with information systems*" [EM3].

Some of these problems were alleviated with SAP. "*At the moment, the integration of information and a single database is critical*" [EM4]; "*There is more transparency*" [EM2]. SAP has proven to be a more intuitive and integrated system at the group level. Although the processes are more complex, the uniformity of SAP has made it "*easier for us to understand the information coming from the other companies and work with it*" [MC1]; furthermore, it enabled the definition of business rules in the system and the dematerialisation of processes (Babaei et al., 2015; Chou et al., 2005; Oliveira & Clegg, 2015; Sheu et al., 2004).

However, new problems emerged, including the lack of agility of the workflow, of some processes and of the system itself – to which, paradoxically, some advantages were also associated (Teittinen et al., 2013): "*It can be good and bad, because the flexibility of the old ERP was used a lot to circumvent the system*" [EM3]. Furthermore, exporting information relating to E2E (end-to-end) processes fell short of expectations. Although SAP was able to verify the relationship between entries and documents throughout the processes, MC had difficulties extracting reports that aggregated and related all the information. Finally, the complexity of the system and the need for a long learning period were emphasised (Granlund & Malmi, 2002). "*It is a system that requires not only knowledge but also continuity. Today, I do things that initially were very difficult for me*" [MC3]. ACAM interviewees identified essentially the same advantages and limitations, but its still early implementation stage had not yet allowed to assess progress in resolving the challenges (Katuu, 2020): "*For example, it is still taking us longer to close the month in the USA*" [MC6]

Interviewees highlighted that SAP brought consistency, reliability and structure to information and processes (Sánchez-Rodríguez & Spraakman, 2012) but remarked a gap in reports, especially for E2E processes, stating that "*SAP does not consolidate multiple companies or different logistics centres within the same company*" [EM1]. They emphasised that BI addresses this gap by linking, consolidating and disseminating management information, ensuring structured, timely and high-quality data, and praised BI for its greater user-friendliness, navigability and accessibility (Chou et al., 2005; Rikhardsson & Yigitbasioglu, 2018; Sánchez-Rodríguez & Spraakman, 2012).

In summary, SAP became the primary information source, providing real-time data, while BI enhanced its analysis. This democratised access beyond MC, with other users receiving daily reports. Information became centralised, updated, reliable and transparent (Sheu et al., 2004). According to one controller: "*If I want to check accounts, I use SAP; if I want a report, I use BI*" [MC5].

BI systems have also made it possible to extend these benefits to companies that still do not have SAP, given their potential for aggregating information (Chou et al., 2005), which has allowed for greater consistency in the information used by the group's different units: "*When we ask something, everyone should have the same answer. That was not the case in the past*" [EM4]. However, the adaptation process was not easy: "*We thought that the adaptation process, particularly in MC, would not be so tough*" [EM4].

2.4.2.2 Change Management

The transformation process at AFL was initiated by the holding company. However, AFL showed great interest due to the clear need for change, as the old system was obsolete, no longer maintained and had no future prospects. Significant openness and active participation resulted from initial sessions that elucidated the need for change. "*People realised that change had to happen*" [EM1]. As the first go-live, AFL was not fully alert to all the challenges, but given the perceptions described above, the employees accepted the change.

Overall, there was little resistance to change, and this was not attributable to a homogenous group of factors or workers. However, the change management process initially did not generate the expected results at AC. Therefore, an additional change process was carried out, in which employees shadowed SAP users in other companies of the group, in order to witness and appreciate, clearly and directly, the change and improvements already achieved. This approach proved effective, providing a deeper understanding and appreciation of the advantages. "*What was interesting was getting them to see what other people were already doing in the units with SAP, and the users showing them the difference*" [EM2]; "*Demonstrating the advantages they had. That was worthwhile*" [EM4].

The PT applied best practices, but restrictions on customising the software limited the immediate adaptation of all processes to the business, leaving gaps to be filled. The processes became more complex and labour-intensive, requiring users to adapt and affecting different areas in different ways (Momoh et al., 2010; Yusuf et al., 2004). "*Our processes had to be changed, most for the better, but not all. Mainly from the point of view of some people, because they have much more work to do, so other people can have better information*" [EM2]. The expected benefits of greater system flexibility and process agility had not yet been fully achieved, and there were still slowdowns and parameterisation errors in some processes, "*which end up requiring more work to correct than if the whole process was carried out manually*" [MC6]. Finally, implementing improvements through SAP customisation was limited due to the need for approval from all affected areas, which entailed more bureaucracy and higher costs than through the previous system.

Other problems included the incorrect definition of approvals workflow, registration errors, incomplete reports and insufficient consideration for Sales unit processes. ACAM also noted the undervaluation of critical gaps, lower service levels due to master data centralisation and lack of immediate support. In addition, time management issues frustrated expectations, with some over-dedication and under-dedication. "*We will need to consolidate and improve what has been implemented and our workflow*" [MC6]. Time management problems prevented meeting some expectations on the go-live: "*Stages like defining processes and validating Business Blueprints took a long time; but others, such as training, took very little time*" [EM2]. Finally, master data maintenance was challenging due to its scale and significant process impact: "*Some units cut corners, but SAP does not allow shortcuts; it requires processes to be done from start to finish. And we [MC], at the end of the chain, were able to see that*" [MC4].

Since AFL was the pioneer, it found itself unable to improve the initial solution after the go-live since the subsequent projects created a shared production environment.

In other words, the phased implementation in other units, combined with the integrative nature of the technology, created a temporary lock-in in one unit, a path dependency phenomenon well studied in organisational literature (e.g., Sydow et al., 2009).

2.4.2.3 Controllers' Tasks and Roles

The implementation of SAP led to changes in the controllers' tasks. BI has made the controllers' work easier and quicker, improving interaction with the system and allowing more indicators to be accessed directly in the system. Some operational activities were transferred to the plants' sites, freeing controllers from manual tasks and allowing them to focus on analysing and reporting (Byrne & Pierce, 2007; Fähndrich, 2023; Rautiainen et al., 2024; Sánchez-Rodríguez & Spraakman, 2012; Wolf et al., 2020).

However, the overall workload increased due to three factors. Firstly, the early stages of system maturity, for example, at ACAM, resulted in an "*increased workload*" [MC5]. Although SAP has generally provided controllers with greater information security and reliability, "*there is still no secure level of trust in the information in SAP; you still have to check the information*" [MC5], as mentioned above; in addition, some expectations about the complete feeding of information into the system have not been fully met, still requiring some information to be collected manually (Granlund & Malmi, 2002; Oesterreich et al., 2019). Secondly, controllers continued to carry out tasks outside MC, such as supporting the purchasing, sales or master data areas – on the one hand increasing controllers' versatility but on the other hand also increasing their workload. Thirdly, it was MC's responsibility to develop BI reports: "*Once you see reporting as an aid to day-to-day management and you trust the tool, you start wanting more reports and information. That is why the [MC] workload tends to increase, but in a good way*" [MC1].

However, despite the changes in the tools and tasks of generating, extracting and analysing information, the role of MC has not changed drastically (Granlund & Malmi, 2002). In AC, the tasks associated with MC "*end up being the same; it is the way we do them that has changed radically, how we go about collecting and preparing the information*" [MC3], involving new content, frequency and concepts (e.g., having the concept of Production Order in the system) (Momoh et al., 2010; Yusuf et al., 2004). Moreover, AC's Executive Management has perceived a loss of quality in the delivery of information for decision-making. There were still challenges, primarily related to the low level of maturity of the ERP implementation, which affected MC's ability to intervene. The expectation was that, with increased maturity of the system over time and new processes (e.g., the customer- or country-orientated profitability analysis process), the quality of the reports would improve (Appelbaum et al., 2017) and the MC's role may be enhanced in time (Rautiainen et al., 2024). "*There is still a lot that MC cannot explain, and it is losing its ability to intervene. There is still much pain in the presentation of results, due to the [low] degree of maturity of the implementation*" [EM4]. However, the complexity and the need for higher skills and adaptation were not seen as negative but as a way forward: "*MC has everything it takes to be much better in 2 or 3 years*" [EM2], after this initial lag (Granlund & Malmi, 2002).

MC has had to adapt, for example, to the new cost or stock valuation models in SAP, where the introduction of standard costs has led to better analyses of production efficiency (Momoh et al., 2010; Yusuf et al., 2004). The controllers' role has become more focused on establishing relationships between data, analysing and reporting it, and oriented towards understanding and monitoring the business (Fähndrich, 2023). "*The role of MC is to understand and monitor the business, an analytical profile*" [EM4].

"*Today, you have a more informed, technically robust profile on MC. The process of extracting information from SAP and analysing it is more demanding*" [EM4]. Despite this, closing the month has become simpler, guaranteeing stability and consistency. Changes in costing assumptions, such as the transition to standard costs, the introduction of three costing views (legal view – company consolidation; profit centre view and group view) and a review of the allocation of indirect costs, have simplified variance analysis. As a result, some interviewees felt that, although complex, "*the technical side of report construction has lost its importance*" [EM1] – which must be understood in relative terms, given the greater importance of critical business analysis and understanding and the simplifications enabled by data integration and centralisation.

Due to new process demands, the workforce expanded in some teams during and after the project. "*Some of the new resources will probably remain because many more things are now done in the system; you need more people with more skills*" [EM3]. So, the major impact was mostly on employees' skills and expected contributions. Concerns also arose about the organisation's historical reliance on MC for decision support, considering the growing importance of IS and the need for MC to adapt to new systems and enhance analytical skills (Rautiainen et al., 2024).

2.5 DISCUSSION

This section explores the broader impacts of MC changes. First, it examines changes in intra-organisational relations and dependencies. Then, it explores the convergence of divergent perspectives and knowledge templates, the ensuing tensions and the importance of hybridity in shaping outcomes, including the gradual move towards the business partner role of MC.

2.5.1 Impacts on Intra-organisational Relations and Dependencies

The integration and structuring of processes in SAP resulted in more complex intra-organisational relations and dependencies between operational areas and MC. Some intra-organisational dependencies increased. MC's dependence increased upon the correct recording of daily transactions regarding lengthier processes and their daily integration into the system. In the opposite direction, other departments became more dependent on MC, since production, handling and dispatch must now wait for the costing process, which became daily. "*Before, costing was done annually and after creating a new article. Now, (…) costing must be carried out constantly*" [MC3].

However, some intra-organisational dependencies decreased. Controllers no longer need to request information files, "*given the integration and standardisation of master data [in SAP]. Having the same article and customer codes facilitated consolidation. Before, we had to collate the information manually*" [MC4]. In the opposite direction, the centralisation of information in the ERP and its democratisation through BI reduced the need to constantly request MC to provide reports. However, "*we are still in a process of cultural change and teaching how to use tools to extract information, without the help of MC*" [MC4].

The above examples signal changes in MC's tasks, roles and skills. Low value-added tasks of manual data input and gathering were reduced. Conversely, there was a rise in the importance of training other organisational areas involved in producing or using MC information. The latter role involved more advanced technical and communication skills, as well as a stronger managerial perspective, particularly at the implementation stage. However, success in this role might actually reduce MC's direct and recurrent involvement and interactions with managers, potentially affecting their influence on decision making if not counterbalanced by other heightened contributions.

Finally, implementing the same ERP at ACAM and AC facilitated collaboration between their MC departments and improved human resource mobility, overcoming the already mentioned difficulties caused by different processes, systems and meanings, hence offering generic, more intangible benefits to the group.

2.5.2 Confluence and Tensions between the PT and the BUs' MC

This case highlighted the overlapping yet distinct perspectives of the PT and the BUs, which can be tracked to their different knowledge templates (Berlinski & Morales, 2024). The PT comprised experienced users, internal functional consultants and external consultants with solid technological knowledge. On the one hand, the PT was highly cognisant of the previous systems' limitations at the local and global levels and focused on achieving global uniformity through ERP and BI, emphasising standardised processes, system-embedded rules (Oliveira & Quinn, 2015) and systems integration. The PT recognised clear benefits of BI in leveraging ERP's capabilities, despite some misalignments with the BU's needs and knowledge, an issue perceived as solvable in time through learning and adaptation.

On the other hand, the BUs appreciated the reliability of the information and the PT's expertise. However, they were less sensitive to some of the prior limitations and were more sensitive to the new limitations and skill gaps. They idealised systems as tools expected to adapt to their functional needs (Berlinski & Morales, 2024), and when confronted with a reality that did not fully match their aspiration, they kept hoping for improvements through learning (at their individual level) and further process improvements (at an organisational level).

Change management strategies acknowledged these differences, employing senior management support, direct benefits visualisation to mitigate resistance (including by shadowing colleagues already using the new systems) and gradually improving the systems – within the constraints of single systems and path dependency effects (Sydow et al., 2009), as experienced within AFL's temporary lock-in in the initially implemented solution. However, the PT's knowledge template tended to become crystallised in the implemented processes and systems, to the detriment of the BUs' knowledge template. This, in turn, limited the BUs' capacity to "know" and "intervene" (Berlinski & Morales, 2024) and, hence, their power (Oliveira & Clegg, 2015).

This case enables a more nuanced reflection on hybridity and the confluence of knowledge templates. The PT's hybridity is collective, combining, in the PT itself, technological and BUs' actors and knowledge; therefore, it is different from Berlinski and Morales' (2024) technologist hybrid actor, with a core IT role albeit with accounting expertise. In a theoretical "hybridity continuum" between "technology" and "business", the PT of our case is a collective entity and has hybridity *within itself* and is overall positioned more towards the business pole than Berlinski and Morales' (2024) hybrid technologist.

This hybrid constitution of the PT was consequential. It allowed for a nuanced balance and for navigating the inherent tensions between standardisation and sensitivity to diverse business needs, recognising that, as quoted above, "*it was not possible to satisfy all the BUs*" [PT2] but striving to incorporate significant business-enabling features into the system. This hybridity and nuanced approach facilitated creating bridges with the BUs' controllers and addressing constraints – even without fully and immediately removing them. Doing so promoted the fluid change (Rautiainen et al., 2024) towards a "business partner" role for controllers, enhancing their contribution to the business and their gradually evolving position within the organisation (Fähndrich, 2023; Rautiainen et al., 2024; Wolf et al., 2020).

2.6 CONCLUSION

This study investigates a project at CA to standardise processes and implement new ERP and BI systems, focusing on changes in MC tasks, roles and skills and using the "people, processes and technology" framework as sensitising lenses to explore the empirics. In terms of technology, the phased roll-out of the new ERP positioned it as the central system at the group level. Additionally, BI was integrated with the ERP, and the data of companies still without SAP was loaded into BI to enable a consolidated database and a single source of information. All processes were revised in line with best practices, striving to strike a balance between the standardisation objective and addressing crucial needs of the BUs – a still ongoing effort. In terms of people, intra-group human resources mobility was facilitated by uniform processes and systems. Primary changes for controllers were concerned with the way controllers work and access information and shifts in tasks across organisation areas. There were also changes in their skills and roles, although not drastic. There was a shift towards information analysis, requiring a more analytical profile, more IS skills (Appelbaum et al., 2017; Oesterreich et al., 2019), more profound business knowledge, the ability to analyse, relate, disseminate and structure information, training skills and additional soft skills. All these changes underpinned a gradual, fluid progress towards the desired evolution from "bean counter" to "business partner" (Byrne & Pierce, 2007; Fähndrich, 2023; Kallunki et al., 2011; Oliveira & Ribeiro, 2023; Rautiainen et al., 2024; Wolf et al., 2020).

The case also reveals that the changes in MC would not have emerged with SAP alone and that BI was essential for aggregating, relating and exploiting data from different systems (Chou et al., 2005; Ehie & Madsen, 2005; Fähndrich, 2023; Momoh et al., 2010; Power & Sharda, 2015; Teittinen et al., 2013). Indeed, the tools used tended to depend on the organisational and decisional level: the operational level used the ERP (or the MES system); the management level used the ERP and primarily BI; and the strategic level only used BI.

The case also highlights more nuanced aspects. The ERP's and BI's integrative potential to automate reporting, disseminate information and promote an increasing demand for richer information and reports changed intra-organisational dependencies between the MC department and other departments, decreasing or increasing those dependencies, and in both directions.

Finally, the case explores the tensions arising from the confluence between different knowledge templates between technologists and the BUs' actors, in particular, controllers (Berlinski & Morales, 2024; Dechow & Mouritsen, 2005). It examines the role of a collective hybrid actor, the PT, in managing the tensions arising from standardisation efforts and local BUs users' needs. The PT accommodated some, but not all, users' requests. Moreover, while characteristics of the adopted solution created a temporary lock-in in the initial configurations (Sydow et al., 2009), there were ongoing efforts to combine the continued expansion of the ERP and BI implementation to other BUs and the accommodation of user requirements not met at the go-live stage. Despite these technologies' infrastructural nature (Yeoh & Koronios, 2010) and the inherent importance of the accepted priorities and requests and adopted options at the implementation stage (Berlinski & Morales, 2024; Dechow & Mouritsen, 2005), this case also shows an ongoing blur between the IS implementation and usage stages (Dechow & Mouritsen, 2005).

This case highlights the pivotal role of a hybrid Project Team (PT) in balancing technological and organisational-wide standardisation with local business needs. It emphasises the PT's collective hybridity, integrating, within itself, diverse knowledge templates and aligning more with business perspectives than the singular hybrid technologist in Berlinski and Morales' (2024) case. This nuanced approach was key to navigating implementation challenges, creating bridges with the BUs' controllers and fostering a gradual shift towards a more value-added "business partner" role for controllers (Goretzki & Messner, 2019; Oliveira & Ribeiro, 2023; Rautiainen et al., 2024; Sánchez-Rodríguez & Spraakman, 2012; Wolf et al., 2020).

This case underscores the importance of understanding complementarities between ERP and BI systems, changing intra-organisational dependencies after their implementation, the importance of hybrid entities and different knowledge templates in managing IS and organisational change, and the gradual shifts in controllers' tasks, roles and skills. These contributions provide important insights for practitioners involved in systems implementation or the MC function, since the issues identified may also emerge in similar processes and organisational contexts. The contributions are also useful for researchers on systems implementation, hybridity, confluence of knowledge templates and controllers' roles.

A limitation of this study stems from the fieldwork having been conducted soon after the implementation of the ERP and BI systems across the group. Given the long-term nature of these systems (Dambrin & Grall, 2023) and the potential lag time in gradual effects on the MC function (Granlund & Malmi, 2002; Rautiainen et al., 2024), a follow-up study within the same organisation would yield valuable insights. Future research may also study the impacts on MC in a company implementing only an ERP but not a BI. Further research may also adopt a broader organisational focus, beyond MC, and whether MC, no longer the primary information supplier, has relinquished control and influence over decisions in other areas (Berlinski & Morales, 2024). Finally, studying the joint mechanisms of power and coalitions underlying organisational changes such as those studied here is also a promising research avenue (Levinthal & Pham, forthcoming).

REFERENCES

Ain, N., Vaia, G., DeLone, W.H., & Waheed, M. (2019). Two decades of research on business intelligence system adoption, utilization and success – A systematic literature review. *Decision Support Systems*, *125*, 113113. https://doi.org/10.1016/j.dss.2019.113113

Appelbaum, D., Kogan, A., Vasarhelyi, M., & Yan, Z. (2017). Impact of business analytics and enterprise systems on managerial accounting. *International Journal of Accounting Information Systems*, *25*, 29–44. https://doi.org/10.1016/j.accinf.2017.03.003

Babaei, M., Gholami, Z., & Altafi, S. (2015). Challenges of enterprise resource planning implementation in Iran large organizations. *Information Systems*, *54*, 15–27. https://doi.org/10.1016/j.is.2015.05.003

Baskerville, R., Myers, M., & Yoo, Y. (2020). Digital first: The ontological reversal and new challenges for information systems research. *MIS Quarterly*, *44*(2), 509–523. https://doi.org/10.25300/MISQ/2020/14418

Berlinski, E., & Morales, J. (2024). Digital technologies and accounting quantification: The emergence of two divergent knowledge templates. *Critical Perspectives in Accounting*, *98*, 102697. https://doi.org/10.1016/j.cpa.2023.102697

Bradley, J. (2008). Management based critical success factors in the implementation of enterprise resource planning systems. *International Journal of Accounting Information Systems*, *9*(3), 175–200. https://doi.org/10.1016/j.accinf.2008.04.001

Byrne, S., & Pierce, B. (2007). Towards a more comprehensive understanding of the roles of management accountants. *European Accounting Review*, *16*(3), 469–498. https://doi.org/10.1080/09638180701507114

Chen, H., Chiang, R.H.L., & Storey, V.C. (2012). Business intelligence and analytics: From big data to big impact. *MIS Quarterly*, *36*(4), 1165–1188.

Chou, D.C., Tripuramallu, H.B., & Chou, A.Y. (2005). BI and ERP integration. *Information Management Computer Security*, *13*(5), 340–349. https://doi.org/10.1108/09685220510627241

Dambrin, C., & Grall, B. (2023). Long-term alterations of ERP systems: Lasting through relinquishment and regeneration. *Qualitative Research in Accounting & Management*, *20*(2), 169–200. https://doi.org/10.1108/QRAM-03-2022-0043

Dechow, N., & Mouritsen, J. (2005). Enterprise resource planning systems, management control and the quest for integration. *Accounting, Organisations & Society*, *30*(7–8), 691–733. https://doi.org/10.1016/j.aos.2004.11.004

Ehie, I. C., & Madsen, M. (2005). Identifying critical issues in enterprise resource planning (ERP) implementation. *Computers in Industry*, *56*(6), 545–557. https://doi.org/10.1016/j.compind.2005.02.006

Fachin, F.F., & Langley, A. (2024). The patterning of interactive organizational identity work. *Organization Studies*, *45*(3), 359–383. https://doi.org/10.1177/01708406231208358

Fähndrich, J. (2023). A literature review on the impact of digitalisation on management control. *Journal of Management Control*, *34*, 9–65. https://doi.org/10.1007/s00187-022-00349-4

Goretzki, L., & Messner, M. (2019). Backstage and frontstage interactions in management accountants' identity work. *Accounting, Organizations and Society*, *74*, 1– 20. https://doi.org/10.1016/j.aos.2018.09.001

Grainger, N., & McKay, J. (2007). ERP: More than just a system. The role of ERP in mergers and acquisitions. *ACIS 2007 Proceedings*, 54.

Granlund, M., & Malmi, T. (2002). Moderate impact of ERPS on management accounting: A lag or permanent outcome? *Management Accounting Research*, *13*(3), 299–321. https://doi.org/10.1006/mare.2002.0189

Kallunki, J.P., Laitinen, E.K., & Silvola, H. (2011). Impact of enterprise resource planning systems on management control systems and firm performance. *International Journal of Accounting Information Systems*, *12*(1), 20–39. https://doi.org/10.1016/j.accinf.2010.02.001

Katuu, S. (2020). Enterprise resource planning: Past, present, and future. *New Review of Information Networking*, *25*(1), 37–46. https://doi.org/10.1080/13614576.2020.1742770

Lambert, C., & Pezet, E., (2011). The making of the management accountant – Becoming the producer of truthful knowledge. *Accounting, Organizations and Society*, *36*(1), 10–30. https://doi.org/10.1016/j.aos.2010.07.005

Leavitt, H.J. (1965). Applied organisational change in industry: Structural, technological and humanistic approaches. In J.G. March (Ed.), *Handbook of Organisation* (pp. 1144–1170). Routledge.

Levinthal, D.A., & Pham, D.N. (Forthcoming). Bringing politics back in: The role of power and coalitions in organizational adaptation. *Organizational Science*. https://doi.org/10.1287/orsc.2022.16995

Lodh, S.C., & Gaffikin, M.J.R. (2003). Implementation of an integrated accounting and cost management system using the SAP system: A field study. *European Accounting Review*, *12*(1), 85–121. https://doi.org/10.1080/0963818031000087899

Momoh, A., Roy, R., & Shehab, E. (2010). Challenges in enterprise resource planning implementation: State-of-the-art. *Business Process Management Journal*, *16*(4), 537–565. https://doi.org/10.1108/14637151011065919

Nofal, M.I., & Yusof, Z.M. (2013). Integration of business intelligence and enterprise resource planning within organizations. *Procedia Technology*, *11*, 658–665. https://doi.org/10.1016/j.protcy.2013.12.242

Oesterreich, T.D., Teuteberg, F., Bensberg, F., & Buscher, G. (2019). The controlling profession in the digital age: Understanding the impact of digitisation on the controller's job roles, skills and competences. *International Journal of Accounting Information Systems*, *35*, 100432. https://doi.org/10.1016/j.accinf.2019.100432

Oliveira, J., & Clegg, S. (2015). Paradoxical puzzles of control and circuits of power. *Qualitative Research in Accounting & Management*, *12*(4), 425–451. https://doi.org/10.1108/QRAM-02-2015-0023

Oliveira, J., & Quinn, M. (2015). Interactions of rules and routines: Re-thinking rules. *Journal of Accounting and Organizational Change*, *11*(4), 503–526. https://doi.org/10.1108/JAOC-11-2013-0095

Oliveira, J. & Ribeiro, P. (2023). Technological developments and new hybrid roles in accounting and finance. In E. Strauss, & M. Quinn (Eds.), *The Routledge Handbook of Accounting Information Systems* (2nd ed., pp. 156–170). Routledge. https://doi.org/10.4324/9781003132943-14

Power, D.J., & Sharda, R. (2015). Business intelligence and analytics. *Wiley Encyclopedia of Management*, 1–4. https://doi.org/10.1002/9781118785317.WEOM070011

Prodan, M., Prodan, A., & Purcarea, A. A. (2015). Three new dimensions to people, process, technology improvement model. In A. Rocha, A.M. Correia, S. Costanzo & L.P. Reis (Eds.), *New Contributions in Information Systems and Technologies: Volume 1* (pp. 481–490). Springer International Publishing.

Rautiainen, A., Scapens, R.W., Järvenpää, M., Auvinen, T, & Sajasalo, P. (2024). Towards fluid role identity of management accountants: A case study of a Finnish bank. *The British Accounting Review*, 101341. https://doi.org/10.1016/j.bar.2024.101341

Rikhardsson, P., & Yigitbasioglu, O. (2018). Business intelligence & analytics in management accounting research: Status and future focus. *International Journal of Accounting Information Systems*, *29*, 37–58. https://doi.org/10.1016/j.accinf.2018.03.001

Sánchez-Rodríguez, C., & Spraakman, G. (2012). ERP systems and management accounting: A multiple case study. *Qualitative Research in Accounting & Management*, *9*(4), 398–414. https://doi.org/10.1108/11766091211282689

Sheu, C., Chae, B., & Yang, C. L. (2004). National differences and ERP implementation: Issues and challenges. *Omega*, *32*(5), 361–371. https://doi.org/10.1016/j.omega.2004.02.001

Spraakman, G., O'Grady, W., Askarany, D. and Akroyd, C. (2018). ERP systems and management accounting: New understandings through "nudging" in qualitative research. *Journal of Accounting & Organizational Change*, *14*(2), 120–137. https://doi.org/10.1108/JAOC-06-2016-0038

Sydow, J., Schreyögg, G., & Koch, J. (2009). Organizational path dependence: Opening the black box. *Academy of Management Review*, *34*(4), 689–709. https://doi.org/10.5465/amr.34.4.zok689

Teittinen, H., Pellinen, J., & Järvenpää, M. (2013). ERP in action – Challenges and benefits for management control in SME context. *International Journal of Accounting Information Systems*, *14*(4), 278–296. https://doi.org/10.1016/j.accinf.2012.03.004

van Roekel, H., & van der Steen, M. (2019). Integration as unrealised ideal of ERP systems – An exploration of complexity resulting from multiple variations of integration. *Qualitative Research in Accounting & Management*, *16*(1), 2–34. https://doi.org/10.1108/QRAM-08-2014-0052

Wolf, T., Kuttner, M., Feldbauer-Durstmüller, B., & Mitter, C. (2020). What we know about management accountants' changing identities and roles – A systematic literature review. *Journal of Accounting & Organizational Change*, *16*(3), 311–347. https://doi.org/10.1108/JAOC-02-2019-0025

Yeoh, W., & Koronios, A. (2010). Critical success factors for business intelligence systems. *Journal of Computer Information Systems*, *50*(3), 23–32. https://doi.org/10.1080/08874417.2010.11645404

Yin, R.K. (2015). *Qualitative Research from Start to Finish*. Guilford Publications.

Yin, R.K. (2018). *Case Study Research and Applications*. Sage.

Yusuf, Y., Gunasekaran, A., & Abthorpe, M.S. (2004). Enterprise information systems project implementation: A case study of ERP in Rolls-Royce. *International Journal of Production Economics*, *87*(3), 251–266. https://doi.org/10.1016/j.ijpe.2003.10.004

3 Firm Internationalization and Activity Impact on Digital Technologies Adoption

Carlos Ribeiro and Sofia Gomes
University Portucalense and REMIT – Research on Economics, Management and Information Technologies, Porto, Portugal

3.1 INTRODUCTION

Enterprise internationalization and technology have experienced rapid evolution in the past two decades, garnering significant attention in numerous academic articles. Most of these studies primarily focus on digitalization's impact on enterprises' internationalization process. However, there may be a reciprocal effect, namely, the impact of enterprise internationalization on digitalization, as discussed by, for example Castagnoli et al. (2021) and Skare and Soriano (2021). These authors also highlight that this reverse causality tends to be under-researched. This study aims to contribute to this body of literature by analyzing seven digital technologies: artificial intelligence (AI), cloud computing, robotics, smart devices, big data analytics (BDA), high-speed infrastructure (HSI), and blockchain. These innovative technologies have begun the Fourth Industrial Revolution (Kumar et al., 2023) and are already being integrated into enterprise systems, Intelligent-Enterprise Resource Planning (ERP). These new systems have advanced capacities to manage BDA (Yu Chung Wang et al., 2022) and already have programmed AI to generate business intelligent dashboards (Gressel et al., 2021). The benefits of BDA, AI, blockchain, HSI (with 5G technology), and cloud computing are already recognized in enterprise systems (Yu Chung Wang et al., 2022). Using smart devices allows for a better alignment of enterprise systems with the physical world, allowing real-time data exchanges (Gao et al., 2011).

In this context, this study aims to explore the complex relationship between adopting digital technology and corporate internationalization. Consequently, this study also aims to investigate how adopting these technologies differs across multiple enterprises' activities. Data for this study was sourced from the Gesis – Leibniz Institute for Social Sciences database, specifically from Flash Eurobarometer 486, published in 2020 by the European Commission (2020). This data encompasses 41 countries and includes information from 16,365 companies across various activities. This sample has already yielded results in the literature (Ferreira et al., 2023).

DOI: 10.1201/9781003461821-3

Seven hypotheses were raised and tested with a linear regression model. Our analysis, broadening the scope beyond manufacturing to encompass all activities, revealed significant variations in the adoption rates of the same seven digital technologies compared to those reported by Ferreira et al. (2023).

The present study provides six specific contributions to the existing literature. Firstly, this is an empirical study with a large sample size of companies (16,365) addressing seven digital technologies. Few empirical studies with such a large sample size cover such a high number of digital technologies simultaneously. Secondly, it aims to measure the impact of a company's internationalization on adopting seven specific digital technologies, thereby contributing to the research stream on reverse causality. In this context, it is evident that this reverse effect positively impacts the chosen digital technologies, albeit with varying degrees of influence. Cloud computing exhibits the highest impact, followed by HSI, smart devices, BDA, robotics, AI, and blockchain. Thirdly, after comparing the technology adoption levels presented by Ferreira et al. (2023), adoption percentages were analyzed across multiple activities. Various firm activities prioritize the adoption of different technologies. This has led to an unintended yet thorough analysis. Fourthly, this study addresses a gap identified in the existing literature through the control variables, as Castagnoli et al. (2021) highlighted. Larger companies exhibit a higher adoption rate of digital technologies than smaller firms, while company age has a notably limited positive effect. Fifthly, after a detailed analysis, we were able to measure the specific impact of the firm's activity under study on adopting the seven digital technologies. The results varied greatly, with some activities even showing negative effects on adopting certain technologies. Sixthly, this study contributes to the existing literature on digital transformation and enterprise systems by offering sector-specific insights essential for businesses seeking to strategically incorporate technology into their domestic or international operations. Intelligent-ERP systems play a pivotal role in this strategic integration, providing a flexible and adaptable platform capable of meeting the distinct technological requirements of each sector.

3.2 LITERATURE REVIEW

3.2.1 Firm Internationalization

Digital technologies such as AI, cloud computing, robotics, smart devices, BDA, HSI, and blockchain have been recognized in current literature as integral to the Industry 4.0 Revolution, as noted by Ghobakhloo and Fathi (2019) and Corallo et al. (2021). This ongoing industrial revolution emphasizes integrated solutions and digitalization, a focus highlighted by Tan et al. (2010). The factors driving this revolution, its challenges, and the consequences of implementing these technologies have been extensively explored, as shown in the literature review by Castagnoli et al. (2021).

In examining the relationship between international business and Industry 4.0, many studies have investigated whether implementing these digital technologies impact a company's internationalization. This impact has been validated, affecting aspects such as the location of internationalization, expansion of business networks, and a faster internationalization process (Castagnoli et al., 2021). However, as mentioned in the introduction, there is a line of study known as the reversal causality research stream, which seeks to understand if a company's internationalization influences technological adoption. These studies often focus on specific industries, with manufacturing being the most examined. They analyze small and medium enterprises separately from multinational companies, preventing direct comparison. The internationalization of companies is also a major factor in globalization, reversing the relationship between companies, cities, and countries (McCann & Acs, 2010). By measuring the impact of companies' internationalization on Industry 4.0, we are also drawing connections between globalization and technological adoption. A research model with seven hypotheses was created to study whether the internationalization of companies impacts the adoption of the seven technologies under investigation. Therefore, we formulate the following hypotheses:

H1:

Firm internationalization has a positive impact on the adoption of a) artificial intelligence; b) cloud computing; c) robotics; d) smart devices; e) big data analytics; f) high-speed infrastructure; and g) blockchain.

Figure 3.1 shows the first research model and the formulated hypotheses.

3.2.2 Firm Activity

The introduction of Industry 4.0 prompted new ways of providing goods and services that disrupt conventional methods. Organizations from different industries have

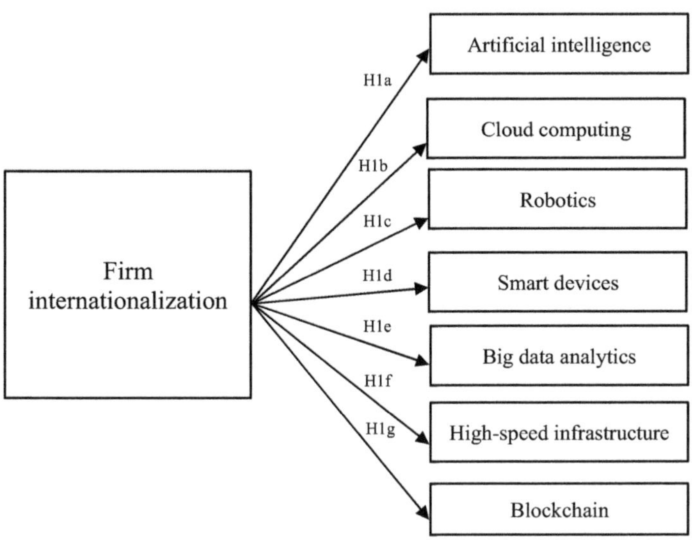

FIGURE 3.1 First research model.

implemented these technologies into their business models to gain a competitive edge and survive in the market (Chakraborty et al., 2021). For instance, the healthcare industry has employed smart devices to track the real-time location of hospital equipment and communicate with its patients offsite through viable data collection and management, the agriculture sector has used smart devices to collect machine metrics and environmental data, and the retail sector has tremendously reduced its operational costs by applying technology that helped in automated checkout, smart inventory management, better supply chain management, and operating fully automatic stores (Kumar et al., 2023). These studies primarily focus on a particular application of technology in a specific industry without comparison to other industries or technologies (e.g., Ben-Daya et al., 2019; Kar & Navin, 2021; Tsiulin et al., 2020). This approach often leads to literature reviews focusing exclusively on specific industries and technologies, potentially overlooking others. Furthermore, some attempts to consolidate information across industries lack comparative data, as the sources are limited to individual industries, as noted in Sharma et al. (2023). By evaluating the impact of company activities on Industry 4.0 technology adoption, this research enriches the literature with a comprehensive view of the activity–technology adoption spectrum. Furthermore, it equips managers with insights into the trends of digital technology adoption occurring within specific industries or across multiple sectors, addressing the common lack of knowledge among managers regarding digitalization trends (Hervé et al., 2020). To streamline the research hypotheses, the focus was on the influence of firm activities on the adopting of the selected technologies. Consequently, a research model (Figure 3.2) comprising the following hypotheses was formulated.

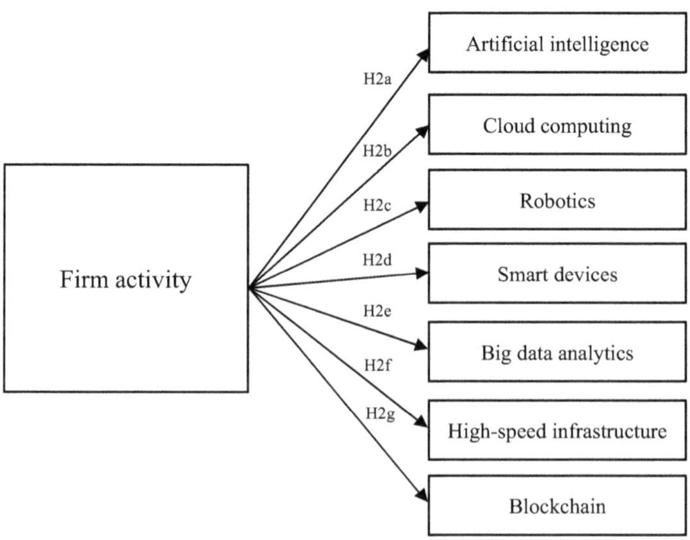

FIGURE 3.2 Second research model.

H2:

Firm activity impacts the adoption of a) artificial intelligence; b) cloud computing; c) robotics; d) smart devices; e) big data analytics; f) high-speed infrastructure; and g) blockchain.

Figure 3.2 shows the second research model and the formulated hypotheses.

3.3 METHOD

3.3.1 Sample and Data

This study aimed to explore the relationship between internationalization and the company's sector of activity with seven different digital technologies: AI, cloud computing, robotics, smart devices, BDA, HSI, and blockchain. Utilizing a dataset sourced from the Gesis – Leibniz Institute for Social Sciences, specifically Flash Eurobarometer 486 published by the European Commission in 2020, the research builds upon a sample that has already proven its academic value, as evidenced by its application in the work of Ferreira et al. (2023). Data collection was conducted via telephone and computer-assisted interviews from February 19, 2020, to May 5, 2020. The sample encompasses a broad geographic scope, covering companies from all 27 European Union member states, as well as Bosnia and Herzegovina, Brazil, Canada, Iceland, Japan, Kosovo, North Macedonia, Norway, Serbia, Turkey, the United Kingdom, and the United States. The database comprises 16,365 companies spanning diverse activities (Table 3.1). The age of companies in the sample ranges from those established in 1850 to those founded in 2020. Regarding the size, 55% of the firms employed between 1 and 9 people; 25% had a workforce of 10 to 49; 14% staffed 50 to 249 individuals; and 5% had over 249 employees. Concerning financial turnover, 33% reported earnings of less than 1 million, 17% declared between 1 and 10 million, and 10% indicated higher revenues. The remainder did not disclose their financial turnover.

3.3.2 Variables

In the Flash Eurobarometer 486 survey, companies responded to question Q9: "Which of the following statements apply to your company?" Participants were allowed to select multiple responses, each being binary: 1 for "yes" and 0 for "no." One of the response options was designed to ascertain whether the company is integrated into a global value chain, specifically stating "It is a part of a global value chain." The answer to this question was used to measure the variable "firm internationalization" in the first study. In the questionnaire, 16,365 companies were assigned a single classification of economic activities (NACE), the official classification of economic activities in the European Union, and indicated in the database variable *nace_a*. From this *nace_a* variable, 16 new binary variables were created, each corresponding to a specific activity. Each company was assigned a binary value of 1 in only one of these sixteen variables, matching their activity classification based on the *nace_a* variable. The answer to this question was used to measure the variable "firm activity" in the second study.

TABLE 3.1
Description of Sample Companies

Country				Industry		No. of Employees		Firm Turnover	
AT – Austria	500	IT – Italy	500	Mining and quarrying	90	1–9	8943	0–100K	1356
BA – Bosnia and Herzegovina	200	JP – Japan	300	Manufacturing	3184	10–49	4133	100–500K	2551
BE – Belgium	500	LT – Lithuania	500	Electricity, gas, steam, and air conditioning supply […]	100	50–249	2328	500K–1M	1458
BG – Bulgaria	500	LU – Luxembourg	200	Water supply, sewerage, and waste management […]	167	250+	816	1–2M	1073
BR – Brazil	344	LV – Latvia	500	Construction	1576	DK/NA	145	2–5M	1066
CA – Canada	500	MK – Makedonia	202	Wholesale and retail trade, repair of motor vehicles […]	4532			5–10M	695
CY – Cyprus (Republic)	201	MT – Malta	201	Transportation and storage	929			10–50M	1376
CZ – Czech Republic	501	NL – The Netherlands	500	Accommodation and food service activities	919			50M+	298
DE – Germany	500	NO – Norway	300	Information and communication	625			DK/NA	6492
DK – Denmark	500	PL – Poland	500	Financial and insurance activities	344				
EE – Estonia	500	PT – Portugal	500	Real estate activities	376				
ES -Spain	502	RO – Romania	500	Professional, scientific, and technical activities	1524				
FI – Finland	501	RS – Serbia	200	Administrative and support service activities	720				
FR – France	503	RS-KM – Kosovo	200	Education	383				
GB – United Kingdom	502	SE – Sweden	500	Human health and social work activities	622				
GR – Greece	500	SI – Slovenia	503	Arts, entertainment, and recreation	274				
HR – Croatia	500	SK – Slovakia	503						
HU – Hungary	500	TR – Turkey	300						
IE – Ireland	500	US – USA	501						
IS – Iceland	201	**Total**	**16,365**		**16,365**		**16,365**		**16,365**

Like Ferreira et al. (2023), control variables were chosen to include firm age and firm size. Firm age was calculated up to the year 2020, beginning from the year of initial registration, as indicated in the database variable *vq1*. Firm size was determined based on the number of employees, as detailed in the database variable *vq2b*.

3.3.2.1 Dependent Variables

The dependent variables in this study were derived from the responses to question Q23 in the Flash Eurobarometer 486 survey, which asked, "Which digital technologies has your company adopted so far?" Companies could select multiple technologies, with each choice being binary: 1 for "yes" and 0 for "no." The specific technologies inquired about were AI, cloud computing, robotics, smart devices, BDA, HSI, and blockchain.

The description of the variables used in this study is shown in Table 3.2.

Table 3.3 contains the description of items of the firm activity variable.

3.3.3 Data Analysis

Firstly, a statistical analysis of the variables was carried out using the IBM SPSS Statistics software version 29.0.2.0. Then, the relationships established in the research models were estimated through linear regression. We aimed to investigate the potential causal relationship between the independent variables, internationalization and economic activity, and adopting digital technologies. Linear regression is a common method for evaluating possible causal relationships between an independent

TABLE 3.2
Description of the Variables

Variables	Description
Independent	
Firm internationalization (first study)	FG1 – Organization is part of a global value chain
Firm activity (second study)	FA – Economic activity by NACE
Dependent	
Artificial intelligence	T1 – Artificial intelligence, e.g., machine learning technologies identifying objects or persons
Cloud computing	T2 – Cloud computing, e.g., storing and processing files or data on remote servers hosted on the internet
Robotics	T3 – Robotics, e.g., robots used to automate processes for example in construction or design
Smart devices	T4 – Smart devices, e.g., smart sensors and smart thermostats
Big data analytics	T5 – Big data analytics, e.g., data mining and predictive analysis
High-speed infrastructure	T6 – High-speed infrastructure
Blockchain	T7 – Blockchain
Control Variables	
Firm age	Length of time (in years) of the company from the initial registration of activity until 2020
Firm size	Number of employees

TABLE 3.3
Items of the Firm Activity Variable

Firm Activity Variables
Mining and quarrying activity
Manufacturing activity
Electricity, gas, steam, and air conditioning supply [...] activity
Water supply, sewerage, and waste management [...] activity
Construction activity
Wholesale and retail trade, repair of motor vehicles [...] activity
Transportation and storage activity
Accommodation and food service activities activity
Information and communication activity
Financial and insurance activities activity
Real estate activities activity
Professional, scientific, and technical activities activity
Administrative and support service activities activity
Education activity
Human health and social work activities activity
Arts, entertainment, and recreation activity

variable and one or more dependent variables, making it suitable for such investigation (Rubinfeld, 2000).

3.4 RESULTS

3.4.1 Descriptive Analysis

The statistical description of the variables used in this study can be found in Tables 3.4 and 3.5. The sample for this study consists of 16,365 companies, of which 10% are part of a global value chain. The most widely adopted technology is cloud computing, at 48%, while HSI is the second most used, with robotics accounting for only 9%. This highlights a distinct pattern in the technological adoption behaviors of multinational manufacturing companies compared to other activities. Robotics was the second most used digital technology in Ferreira et al.'s (2023) study.

Given this significant difference, providing a disaggregated demonstration by activity became important. It was also observed that international companies have a much higher level of adopting of digital technology, except for blockchain, where their adoption rate is lower than that of non-international companies.

In this study, the most represented activities are wholesale and retail trade, repair of motor vehicles, manufacturing, and construction. The information and communication activity is the most prolific adopter of AI, cloud computing, BDA, and HSI. Manufacturing activity has led to the adoption of robotics. The electricity, gas, steam, and air conditioning supply activity is the primary adopter of smart devices. Interestingly, the education sector emerges as the top adopter of blockchain technology. Conversely, the electricity, gas, steam, and air conditioning supply activity shows the lowest levels of technological adoption, with 45% of the surveyed companies indicating they do not utilize any of these digital technologies.

TABLE 3.4
Descriptive Data on Internationalization Using Digital Technologies Adopted

	Total (%)	Non-international Organization (%)	International Organization (%)
Total	100	90	10
Artificial intelligence	8	6	2
		7	17
Cloud computing	48	41	7
		46	70
Robotics	9	7	2
		7	21
Smart devices	28	23	4
		26	45
Big data analytics	14	12	3
		13	30
High-speed infrastructure	34	29	5
		32	52
Blockchain	27	24	2
		27	26
None of the above	28	25	2
		28	26

3.4.2 Research Models Testing

3.4.2.1 Research Model 1

Table 3.6 contains the results of the multiple linear regression, in which the independent variable is firm internationalization and the dependent variables are adopting digital technologies. The control variables are firm age and firm size. Firm internationalization positively impacts the adoption of all technologies, confirming all model hypotheses. However, the impact varies considerably across these technologies. Cloud computing exhibits the highest impact ($\beta = 0.246$), followed by HSI ($\beta = 0.207$), smart devices ($\beta = 0.189$), BDA ($\beta = 0.171$), robotics ($\beta = 0.133$), and AI ($\beta = 0.101$). Blockchain ($\beta = 0.039$) shows the lowest positive causality impact. The hypotheses H1a to H1g were confirmed.

Although not the primary focus of this study, the control variable of firm size demonstrates a more significant positive impact than anticipated by the authors. Larger firms tend to have a higher adoption rate. It's important to emphasize that the previous statement provides valuable insights into addressing a specific gap highlighted in a systematic literature review, which concerns the lack of a clear understanding of which type of firms, whether small, medium, or larger enterprises, exhibits a greater adoption of digital technology (Castagnoli et al., 2021). The control variable of firm age also has a very scarce positive impact on most technologies' adoption.

Table 3.7 contains the results of the multiple linear regression, in which the independent variable is firm activity and the dependent variables are adopting digital technologies. The statistically significant threshold was set at a P-Value

TABLE 3.5
Descriptive Data on Economic Activity Using Digital Technologies Adopted

Total (%)		Artificial Intelligence (%)		Cloud Computing (%)		Robotics (%)		Smart Devices (%)		Big Data Analytics (%)		High-Speed Infrastructure (%)		Blockchain (%)		None of the Above (%)	
Mining and quarrying	1	0.04	8	0.25	46	0.05	9	0.24	43	0.09	16	0.12	22	0.18	33	0.14	26
Manufacturing	19	1.61	8	8.81	45	3.76	19	6.43	33	2.68	14	6.06	31	5.42	28	5.20	27
Electricity, gas, steam, and air conditioning supply […]	1	0.06	10	0.30	49	0.08	13	0.28	46	0.13	21	0.27	45	0.09	15	0.27	45
Water supply, sewerage, and waste management […]	1	0.04	4	0.46	46	0.07	7	0.33	32	0.21	20	0.21	20	0.35	34	0.29	28
Construction	10	0.42	4	4.03	42	0.44	5	2.41	25	0.66	7	2.55	27	2.52	26	2.57	27
Wholesale and retail trade, repair of motor vehicles […]	28	1.59	6	12.13	44	1.75	6	6.94	25	3.97	14	9.17	33	7.39	27	7.30	26
Transportation and storage	6	0.29	5	2.27	40	0.14	2	1.56	28	0.68	12	1.61	28	1.46	26	1.59	28
Accommodation and food service activities	6	0.29	5	2.07	37	0.18	3	1.39	25	0.61	11	1.62	29	1.78	32	1.49	27
Information and communication	4	0.96	25	2.85	75	0.37	10	1.31	34	1.27	33	2.13	56	0.81	21	1.20	32
Financial and insurance activities	2	0.24	12	1.41	67	0.23	11	0.65	31	0.53	25	1.12	53	0.31	15	0.66	31
Real estate activities	2	0.18	8	1.22	53	0.08	3	0.70	31	0.35	15	0.91	40	0.56	24	0.63	27
Professional, scientific, and technical activities	9	1.09	12	5.68	61	0.76	8	2.17	23	1.57	17	3.68	40	1.81	19	2.99	32
Administrative and support service activities	4	0.34	8	2.36	54	0.24	5	1.08	24	0.65	15	1.55	35	1.00	23	1.19	27
Education	2	0.13	5	1.25	53	0.15	6	0.65	28	0.29	12	0.81	35	0.93	40	0.60	26
Human health and social work activities	4	0.22	6	1.94	51	0.20	5	1.11	29	0.59	15	1.29	34	1.45	38	1.04	27
Arts, entertainment, and recreation	2	0.13	8	0.85	51	0.08	5	0.53	32	0.19	11	0.62	37	0.59	35	0.48	28

TABLE 3.6
Linear Regression Results of Model 1

	AI			Cloud Computing			Robotics			Smart Devices			BDA			High-Speed Infrastructure			Blockchain		
	Coeff.	T Statis-tics	P-Value	Coeff.	T Statis-tics	P-Value	Coeff.	T Statis-tics	P-Value	Coeff.	T Statis-tics	P-Value	Coeff.	T Statis-tics	P-Value	Coeff.	T Statis-tics	P-Value	Coeff.	T Statis-tics	P-Value
Firm internation-alization	0.101	14.459	0.000	0.246	18.754	0.000	0.133	18.037	0.000	0.189	16.037	0.000	0.171	18.511	0.000	0.207	16.605	0.000	0.039	8.196	0.000
Firm age	0.000	2.553	0.011	0.001	5.708	0.000	0.001	9.505	0.000	0.001	6.964	0.000	0.001	6.714	0.000	0.000	–0.051	0.960	–0.001	–6.384	0.000
Firm size	0.074	3.277	0.001	0.068	1.860	0.065	0.086	3.649	0.000	0.118	3.484	0.000	0.138	4.683	0.000	0.000	2.944	0.003	0.022	1.102	0.272

TABLE 3.7
Multiple Regression Results of Model 2

	AI			Cloud Computing			Robotics			Smart Devices			BDA			High-Speed Infrastructure			Blockchain		
	Coeff.	T Statis-tics	P-Value	Coeff.	T Statis-tics	P-Value	Coeff.	T Statis-tics	P-Value	Coeff.	T Statis-tics	P-Value	Coeff.	T Statis-tics	P-Value	Coeff.	T Statis-tics	P-Value	Coeff.	T Statis-tics	P-Value
Mining and quarrying	0.020	0.722	0.470	0.018	0.335	0.738	0.026	0.875	0.382	0.183	3.845	0.000	0.012	0.327	0.744	−0.109	−2.181	0.029	0.012	0.644	0.520
Manufacturing	0.025	4.168	0.000	0.015	1.308	0.191	0.130	20.507	0.000	0.080	7.757	0.000	−0.006	−0.688	0.491	−0.019	−1.793	0.073	0.005	1.098	0.272
Electricity, gas, steam, and air conditioning supply […]	0.042	1.597	0.110	0.052	1.045	0.296	0.067	2.403	0.016	0.209	4.639	0.000	0.067	1.890	0.059	0.119	2.514	0.012	−0.002	−0.123	0.902
Water supply, sewerage, and waste management […]	−0.022	−1.046	0.295	0.017	0.441	0.659	0.009	0.394	0.693	0.073	2.067	0.039	0.060	2.191	0.028	−0.127	−3.452	0.001	−0.014	−1.014	0.311
Construction	−0.014	−1.797	0.072	−0.020	−1.379	0.168	−0.018	−2.198	0.028	−0.000	−0.002	0.998	−0.075	−7.349	0.000	−0.066	−4.800	0.000	−0.015	−2.891	0.004
Wholesale and retail trade, repair of motor vehicles […]	excluded			excluded			excluded			excluded			excluded			excluded			excluded		
Transportation and storage	−0.007	−0.740	0.460	−0.039	−2.179	0.029	−0.039	−3.902	0.000	0.025	1.549	0.121	−0.024	−1.907	0.056	−0.048	−2.838	0.005	−0.003	−0.491	0.624
Accommodation and food service activities	−0.005	−0.564	0.573	−0.069	−3.881	0.000	−0.031	−3.090	0.002	−0.003	−0.159	0.874	−0.035	−2.745	0.006	−0.043	−2.515	0.012	−0.015	−2.294	0.022
Information and communication	0.194	17.269	0.000	0.308	14.644	0.000	0.033	2.790	0.005	0.093	4.901	0.000	0.189	12.735	0.000	0.227	11.379	0.000	0.053	6.908	0.000
Financial and insurance activities	0.059	3.994	0.000	0.234	8.482	0.000	0.044	2.882	0.004	0.057	2.303	0.021	0.109	5.617	0.000	0.204	7.784	0.000	0.023	2.307	0.021
Real estate activities	0.022	1.574	0.115	0.094	3.555	0.000	−0.029	−1.952	0.051	0.055	2.304	0.021	0.011	0.579	0.563	0.065	2.598	0.009	−0.006	−0.587	0.557
Professional, scientific, and technical activities	0.060	7.694	0.000	0.172	11.816	0.000	0.018	2.220	0.026	−0.018	−1.341	0.180	0.025	2.443	0.015	0.065	4.664	0.000	0.006	1.230	0.219
Administrative and support service activities	0.019	1.783	0.075	0.100	5.038	0.000	−0.009	−0.832	0.405	−0.006	−0.347	0.728	0.005	0.371	0.711	0.022	1.160	0.246	0.001	0.156	0.876
Education	−0.003	−0.197	0.844	0.095	3.613	0.000	−0.001	−0.045	0.964	0.029	1.209	0.227	−0.021	−1.117	0.264	0.016	0.653	0.514	−0.009	−0.918	0.359
Human health and social work activities	0.000	0.026	0.980	0.072	3.404	0.001	−0.012	−1.012	0.311	0.040	2.114	0.035	0.011	0.733	0.464	0.008	0.412	0.680	−0.002	−0.219	0.827
Arts, entertainment, and recreation	0.023	1.389	0.165	0.069	2.263	0.024	−0.016	−0.930	0.352	0.067	2.408	0.016	−0.030	−1.397	0.162	0.038	1.292	0.197	−0.014	−1.258	0.208

less or equal to 0.05. Therefore, results with higher P-Values were considered not statistically significant. The mining and quarrying activity positively impacts smart devices (β = 0.183) but negatively impacts HSI adoption (β = −0.109). The manufacturing activity has a positive impact on robotics (β = 0.130), smart devices (β = 0.080), and AI (β = 0.025) adoption. The electricity, gas, steam, and air conditioning supply [...] activity has a positive impact on smart devices (β = 0.209), HSI (β = 0.119), and robotics (β = 0.067) adoption. The water supply, sewerage, and waste management [...] activity has a positive impact on smart devices (β = 0.073) and BDA (β = 0.060) but a negative impact on HSI (β = −0.127) adoption. The construction activity negatively impacts robotics (β = −0.018), BDA (β = −0.075), HSI (β = −0.066), and blockchain (β = −0.015) adoption. Transportation and storage activity negatively impacts the adoption of robotics (β = −0.039) and HSI (β = −0.048). Accommodation and food service activities activity negatively impacts cloud computing (β = −0.069), robotics (β = −0.031), big data analytics (β = −0.035), HSI (β = −0.043), and blockchain (β = −0.015) adoption. Information and communication activity has different positive impacts on the adoption of the seven selected digital technologies: AI (β = 0.194), cloud computing (β = 0.308), robotics (β = 0.033), smart devices (β = 0.093), big data analytics (β = 0.189), HSI (β = 0.227), and blockchain (β = 0.05). Financial and insurance activities activity also has seven different positive impacts on the adoption: AI (β = 0.059), cloud computing (β = 0.234), robotics (β = 0.044), smart devices (β = 0.057), big data analytics (β = 0.109), HSI (β = 0.204), and blockchain (β = 0.023). Real estate activity has a positive impact on cloud computing (β = 0.094), HSI (β = 0.065), and smart devices (β = 0.051) adoption. Professional, scientific, and technical activities activity has a positive impact on cloud computing (β = 0.172), HSI (β = 0.065), AI (β = 0.060), big data analytics (β = 0.025), and robotics (β = 0.018) adoption. Administrative and support service activities and education activity both have a positive impact on cloud computing (β = 0.100; β = 0.095) adoption. Human health and social work activities activity positively impacts on cloud computing (β = 0.072) and smart devices (β = 0.40) adoption. Arts, entertainment, and recreation activity has a positive impact on cloud computing (β = 0.069) and smart devices (β = 0.067) adoption. Wholesale and retail trade [...] activity was automatically excluded from the research model by the software, as it doesn't bring additional significant information. Thus, hypotheses H2a to H2g were confirmed.

3.5 DISCUSSION AND IMPLICATIONS

The study's results reveal intricate dynamics, indicating the varying impact of firm internationalization on different technologies. Cloud computing's significant impact (β = 0.246) suggests its vital role in global operations, offering scalable and accessible data management crucial for firms operating internationally, underscoring its importance in Intelligent-ERP systems. HSI and smart devices, with impacts of β = 0.207 and β = 0.189, respectively, are also highly relevant, likely due to their contribution to enhancing global connectivity and operational efficiency, ensuring communication and real-time data exchange between the physical world and Intelligent-ERP systems. BDA, robotics, and AI show notable but lesser impacts (β = 0.171, β = 0.133, and

$\beta = 0.101$, respectively). This could be because these technologies, while important, may be integrated more gradually into international business practices or their applications might be more specific compared to the broad utility of cloud computing and HSI. Blockchain, despite its lower impact ($\beta = 0.039$), is still positively influenced by internationalization. Its specific applications in areas like secure transactions and supply chain management might be valuable but are not as universally required as other technologies. Overall, the confirmed hypotheses H1 collectively demonstrate that while internationalization invariably drives technology adoption, the extent of this influence is contingent on the technology's perceived utility and alignment with the firm's international operations and strategy. This nuanced understanding is crucial for firms strategically implementing technology in their international business models. Additionally, various industries are adapting to the digital era with unique patterns of technology adoption, as indicated by the confirmation of hypotheses H2. For example, in mining and quarrying, the positive inclination toward smart devices suggests an emphasis on technology that enhances operational efficiency. This is contrasted by a lesser focus on high-speed infrastructure, possibly due to the industry's operational characteristics or existing infrastructure challenges. Conversely, manufacturing shows a strong movement toward automation, evidenced in the positive impact on robotics, smart devices, and AI, aligning with global trends toward efficient, tech-driven production processes. In energy-related sectors like electricity, gas, steam, and air conditioning supply, adopting smart devices, HSI, and robotics indicates a prioritization of efficient control systems, reliable communication networks, and automated operations, which are essential for managing complex and critical energy distribution and maintenance tasks. This trend is echoed in water supply, sewerage, and waste management sectors, where smart devices and BDA are increasingly relied upon for operational management. However, the negative impact on HSI adoption in these sectors might reflect different priorities, budget constraints, or challenges in integrating such technologies into existing systems. These findings offer a detailed understanding of how technological adoption plays out across different industries.

3.5.1 Managerial and Practical Implications

In practical terms, the study suggests that companies, especially those engaged in international operations, should assess the added value of adopting certain technologies within their Intelligent-ERP systems. Since internationalization is a driving force for technology adoption, companies not yet utilizing technologies such as cloud computing and HSI might be at a competitive disadvantage. Integrating these technologies into Intelligent-ERP systems can significantly enhance global operational capabilities and data management, which are crucial for firms operating across borders. Moreover, the study's activity-specific analysis indicates that adopting specific technologies can provide distinct advantages within certain sectors. For instance, companies in mining and quarrying might find substantial value in incorporating smart devices. Manufacturing firms could benefit from integrating robotics and AI. For electricity, gas, steam, and air conditioning suppliers, adopting of smart devices and HSI could be key to optimizing energy distribution and resource management. The study also highlights the potential for BDA within sectors like water supply, sewerage, and waste management. Furthermore, the broad applicability of cloud computing across various sectors suggests that its adoption can provide scalability, enhanced data security, and global accessibility, supporting firms in maintaining their competitive edge. The emphasis on HSI underscores its importance in supporting real-time data needs, ensuring efficient and seamless operations across diverse geographical locations. In conclusion, the study's results advocate for a strategic approach to technology adoption tailored to different sectors' specific needs and challenges. Technology adoption in companies is affected by multiple factors, including the popularity of the technology within the industry (Gibbs et al., 2007).

3.6 CONCLUSION

This study highlights the intricate relationship between a firm's internationalization and its adoption of digital technologies. It underscores the influence of global business activities on adopting technologies like AI, cloud computing, and BDA. The findings suggest that internationalization is not just a consequence but also a catalyst for technological advancement. This research contributes significantly to our understanding of how digital technologies are integrated into global business strategies. It also sheds light on how this relationship varies across different business activities – technology adoption is not a one-size-fits-all approach. For instance, the adoption of smart devices in mining and quarrying, robotics in manufacturing, and cloud computing in sectors like finance and education demonstrates these technologies' unique value to each field. This sector-specific insight is crucial for businesses aiming to strategically integrate technology into their operations, domestically or internationally, with Intelligent-ERP systems playing a pivotal role in this strategic integration by providing a flexible and adaptable platform that can accommodate the unique technology needs of each sector. The study's limitations include reliance on secondary data from a single database, limiting the scope of the indicators collected by this database. The data, dating back to 2020, encompasses companies from multiple countries. Recognizing this, future research could benefit from incorporating more recent data to enhance the study's relevance and applicability. Additionally, future research could incorporate different technologies or technology topics such as cybersecurity or DevOps. Future studies can also consider explaining why different business activities tend to adopt different technologies and how the adoption process varies (e.g., How do mining and quarrying firms benefit from smart devices? How do water supply, sewerage, and waste management firms benefit from BDA? How do arts, entertainment, and recreation firms benefit from cloud computing? How does the adoption process of cloud computing vary across industries? What are the best adoption practices for robotics in manufacturing firms? What are the best adoption practices for AI in information and technologies firms?).

REFERENCES

Ben-Daya, M., Hassini, E., & Bahroun, Z. (2019). Internet of things and supply chain management: A literature review. *International Journal of Production Research*, 57(15–16), 4719–4742.

Castagnoli, R., Büchi, G., Coeurderoy, R., & Cugno, M. (2021). Evolution of industry 4.0 and international business: A systematic literature review and a research agenda. *European Management Journal*, 40(4). 572–589.

Chakraborty, S., Bhatt, V., Chakravorty, T., & Chakraborty, K. (2021). Analysis of digital technologies as antecedent to care service transparency and orchestration. *Technology in Society*, 65, 101568.

Corallo, A., Crespino, A. M., Vecchio, V. D., Lazoi, M., & Marra, M. (2021). Understanding and defining dark data for the manufacturing industry. *IEEE Transactions on Engineering Management*, 70, 1–13.

European Commission (2020). Flash Eurobarometer 486 (SMEs, start-ups, scale-ups and entrepreneurship). Retrieved from Brussels: https://search.gesis.org/research_data/ZA7637?doi=10.4232/1.1363

Ferreira, J. J., Lopes, J. M., Gomes, S., & Rammal, H. G. (2023). Industry 4.0 implementation: Environmental and social sustainability in manufacturing multinational enterprises. *Journal of Cleaner Production*, 404, 136841.

Gao, F., Zaremba, M., Sami Bhiri, & Derguerch, Wassim (2011). Extending BPMN 2.0 with sensor and smart device business functions. In *2011 IEEE 20th International Workshops on Enabling Technologies: Infrastructure for Collaborative Enterprises* (pp. 297–302). IEEE.

Ghobakhloo, M., & Fathi, M. (2019). Corporate survival in Industry 4.0 era: The enabling role of lean-digitized manufacturing. *Journal of Manufacturing Technology Management*, 31, 1–30.

Gibbs, S., Sequeira, J., & White, M. M. (2007). Social networks and technology adoption in small business. *International Journal of Globalisation and Small Business*, 2(1), 66.

Gressel, S., Pauleen, D. J., & Taskin, N. (2021). *Management Decision-making, Big Data, and Analytics*. SAGE.

Hervé, A., Schmitt, C., & Baldegger, R. (2020). Internationalization and digitalization: Applying digital technologies to the internationalization process of small and medium-sized enterprises. *Technology Innovation Management Review*, 10(7), 28–40.

Kar, A. K., & Navin, L. (2021). Diffusion of blockchain in insurance industry: An analysis through the review of academic and trade literature. *Telematics and Informatics*, 58, 101532.

Kumar, A., Dhingra, S., & Falwadiya, Himanshu (2023). Adoption of internet of things: A systematic literature review and future research agenda. *International Journal of Consumer Studies*, 47(6), 2553–2582.

McCann, P., & Acs, Z. J. (2010). Globalization: Countries, cities and multinationals. *Regional Studies*, 45(1), 17–32.

Rubinfeld, D. L. (2000). Reference guide on multiple regression. *Reference Manual on Scientific Evidence*, 179, 425–469.

Sharma, S., Malik, A., Sharma, C., Batra, I., Kaswan, Mahender Singh, & Garza-Reyes, Jose Arturo (2023). Adoption of industry 4.0 in different sectors: A structural review using natural language processing. *International Journal on Interactive Design and Manufacturing*, 1–23.

Skare, M., & Soriano, D. R. (2021). How globalization is changing digital technology adoption: An international perspective. *Journal of Innovation & Knowledge*, 6(4), 222–233.

Tan, W., Xu, Y., Xu, W., Xu, L., Zhao, X., Wang, L., & Fu, L. (2010). A methodology toward manufacturing grid-based virtual enterprise operation platform. *Enterprise Information Systems*, 4, 283–309.

Tsiulin, S., Reinau, K. H., Hilmola, O.-P., Goryaev, N., & Karam, A. (2020). Blockchain-based applications in shipping and port management: A literature review towards defining key conceptual frameworks. *Review of International Business and Strategy*, 30(2), 201–224.

Yu Chung Wang, W., Pauleen, D., & Taskin, N. (2022). Enterprise systems, emerging technologies, and the data-driven knowledge organisation. *Knowledge Management Research & Practice*, 20(1), 1–13.

4 Recommendation Systems in E-commerce

Link Prediction in Multilayer Bipartite Networks

Luís Ramoa and Pedro Campos
University of Porto, Porto, Portugal

4.1 INTRODUCTION

Incorporating recommendation systems (RS) in e-commerce (EC) aligns with the broader goals of digital transformation by enhancing customer experiences, promoting data-driven decision-making, fostering agile strategies, optimizing operations, fostering innovation, and embracing a customer-centric and adaptive approach. By recommending products based on predicted links, EC platforms can enhance customer engagement and increase retention, leading to higher customer lifetime value. Yang et al. (2024) address Exclusive Enterprise Social Networks (EESNs) to expand sales and provide after-sale services by formulating the social behaviour analysis in EESNs as a link prediction (LP) problem in heterogeneous social networks.

In inventory management, LP can help anticipate connections between products and customer segments aids in forecasting demand and popularity. EC businesses can optimize inventory levels, reduce overstock or stockouts, and improve overall supply chain efficiency. Kosasih (2023) investigates the development of trustworthy artificial intelligence as a data-driven approach to predict hidden dependencies in supply chain with impact on inventory management. LP can also help identify potential links between products and enable the recommendation of related or complementary items with business impact on cross-selling and up-selling strategies, leading to increased average order value and maximizing revenue per customer. Behera et al. (2020) described a model for providing real-time personalized marketing information on recommended products to online and offline shoppers using a combination of sales strategies: up-selling, cross-selling, best-in-class up-selling, and meeting needs. Mathews et al. (2021) use LP techniques to handle tax evasion, by automating the process of detecting circular trading. The model helps to target on a set of plausible evaders and take appropriate preventive measures to avoid fraud.

In marketing campaign optimization, LP can be used to understand how the relationship between marketing efforts and customer responses helps optimize targeted campaigns, allowing to allocate marketing resources more effectively, ensuring a higher return on investment for promotional activities. Vikatos et al. (2020) introduce

DOI: 10.1201/9781003461821-4

a method to enhance marketing campaign targeting by employing bridge prediction in communities within multilayer online social networks. The strategy involves identifying nodes with high brand loyalty and top-ranking nodes participating in bridges. Experimental results, using a X and Foursquare dataset, validate their approach, as this methodology benefits recommendation systems and marketers aiming to leverage social influence for effective marketing campaigns.

This chapter explores EC and RS and the general research question is going to be based on the recommendation associated with purchases in EC: How to recommend products from the same category or from different categories, suggesting cross-selling and up-selling logics based on this logic of clarity and explainability?

Our methodological proposal involves the use of Multilayer Bipartite Networks (B-MLNs) to understand and establish the connections between customers and products in the form of LP.

4.2 LITERATURE REVIEW

4.2.1 E-COMMERCE

According to Alamdari et al. (2020) and Turban et al. (2018) the term EC refers to any form of business that is conducted or transferred electronically. The growth rate of EC has been increasing over the years owing to the gradual decrease in the costs of the Internet (He, 2021). It is estimated that in 2020 alone more than two billion people made purchases of goods or services online, translating into sales exceeding 4.2 trillion dollars worldwide, which corresponds to a growth in EC sales of over 25% compared to 2019 (Coppola, 2021). The emergence of the COVID-19 pandemic has only served to accelerate this shift of consumers towards the digital world.

On one hand, EC presents numerous advantages, at both the organizational and customer levels, such as global reach; EC operates 24 hours a day, 7 days a week; access to detailed information; non-cash payment; advertising/marketing – EC contributes to greater efficiency in product/service marketing management through the use of recommendation systems for sales personalization and customer loyalty; and the ability to quickly compare prices for the same product across different e-shops (Alamdari et al., 2020; Išoraitė & Miniotienė, 2018; Vadwala & Vadwala, 2017). On the other hand, EC also has disadvantages, such as authentication issues; increased insecurity; the ease of price errors; lack of privacy; delivery time; refunds, among others (Išoraitė & Miniotienė, 2018).

Two important concepts to mention are cross-selling and up-selling strategies, which can help increase revenues from current customers. Cross-selling occurs when a company sells different products to its customers (Blattberg et al., 2008). Up-selling is defined as a strategy of asking a customer if they would like to purchase an additional item or a similar item from the same category but of higher value (Lesage et al., 2020).

4.2.2 RECOMMENDATION SYSTEMS

RS are information filtering systems that aim to predict user preferences, with the goal of recommending a personalized list of items that may be of interest to a particular user (Almonte et al., 2021; Alrumiah & Hadwan, 2021; Lakshmi & Bhavani, 2021). They are typically employed in scenarios, websites, or applications where there is a wide variety of products/items and the search and selection of those become complicated for the user, as there are numerous options to choose from (Almonte et al., 2021; Lakshmi & Bhavani, 2021). The effective implementation of intelligent RS has made it possible to acquire insights that provide businesses with valuable information to understand user behaviour, thereby increasing revenues and fostering customer loyalty or attracting new customers, stimulating the exponential growth of various EC companies (Alrumiah & Hadwan, 2021; Ballestar, 2021). For example, with effective RS, Netflix increased views by 75%, Amazon experienced a 35% revenue increase, and YouTube managed to increase views by 60% (Lakshmi & Bhavani, 2021).

To be efficient and effective, RS need to collect and analyse a diverse set of data, whether it's explicit or implicit. Explicit data corresponds to information provided directly by users. Implicit data pertains to information collected implicitly, i.e., information obtained from monitoring user behaviour, such as queries from past searches; previous purchases; items purchased on EC sites; and places where users clicked on the sites (Alamdari et al., 2020; Almonte et al., 2021; Lakshmi & Bhavani, 2021).

4.2.2.1 Relevant Aspects of Recommendation Systems

There are several relevant aspects to mention about RS that often correspond to many of their problems and that may also be useful when analysing their performance (Alamdari et al., 2020; Fayyaz et al., 2020; Lakshmi & Bhavani, 2021): **Precision –** The greater the system's precision, the better the quality of the recommendations, and consequently, it should lead to an increase in sales and greater customer loyalty in EC. **Scalability –** A scalable system can handle growing demands without compromising efficiency or overloading the system. **Security –** RS collect personal information from users, necessitating high security standards to protect valuable data. **Sparsity –** The reduced rating of items by users (there are usually few user/item pairs) translates into a scarcity of data to "feed" the RS (Fayyaz et al., 2020; Lakshmi & Bhavani, 2021). Therefore, this impacts recommendation quality for users who don't provide ratings or feedback (Fayyaz et al., 2020). **Cold Start –** Along with sparsity, cold start is one of the most common and relevant problems of RS. Cold start occurs whenever a new product is introduced on an EC site (cold start products) or when a new user creates an account for the first time (cold start users). In these cases, as there is no information and interactions about them, the RS have no reference or history available to base the recommendations to the new user or will not know when to recommend an ad related to a new product (Fayyaz et al., 2020).

4.2.2.2 Types of Recommendation Systems

The types of RS are not fixed, and the number of existing RS varies according to the author and the degree of importance or usage of RS. Thus, depending on how they generate personalized recommendations, RS can be represented according to the taxonomy presented in Figure 4.1 (Alamdari et al., 2020; Lakshmi & Bhavani, 2021).

Collaborative Filtering Systems (CFs) are prominent RS in EC, leveraging customer ratings and reviews for accurate suggestions. User-based CF predicts item ratings by identifying similar users and obtaining their ratings. Item-based CF focuses on user ratings for similar items, often providing more accurate recommendations, particularly in high-user scenarios. Memory-based algorithms like Pearson correlation coefficient and model-based algorithms like matrix factorization are employed

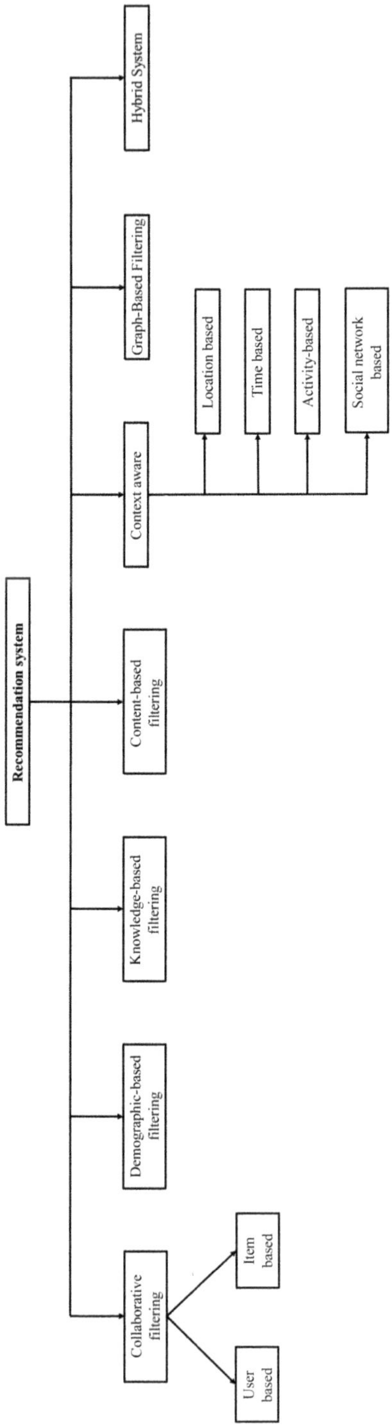

FIGURE 4.1 Classification of recommendation systems.

(Source: Own elaboration.)

to calculate user similarities (Alamdari et al., 2020; Almonte et al., 2021; Das et al., 2021; Fayyaz et al., 2020; Fogelman-Soulié et al., 2020; Ricci et al., 2015; Sarwar et al., 2001). Despite advantages like relevance and quality, CF faces challenges such as cold start issues and data sparsity, impacting scalability and response time. Amazon.com predominantly uses item-based CF due to its simplicity, scalability, and performance. Moreover, the integration of machine learning techniques is growing rapidly (Alamdari et al., 2020; Lakshmi & Bhavani, 2021; Smith & Linden, 2017).

Demographic-Based Filtering Systems utilize user profiles, considering factors like location and age for personalized recommendations. While advantageous for addressing scalability and cold start issues, they face drawbacks like privacy concerns and reliance on user interests (Alamdari et al., 2020; Almonte et al., 2021; Fayyaz et al., 2020). Knowledge-Based Systems (KBs) utilize explicit knowledge about preferences to offer personalized recommendations. KB overcomes cold start problems but has downsides like high operating costs and low scalability (Alamdari et al., 2020; Almonte et al., 2021; Fayyaz et al., 2020). Content-Based Filtering Systems (CBFs) construct user profiles based on preferences, recommending items similar to the user's liking. While independent and scalable, CBF demands intensive domain knowledge and may compromise privacy (Alamdari et al., 2020; Cerqueira et al., 2016; Fayyaz et al., 2020; Lakshmi & Bhavani, 2021). Context-Aware Systems use user context to enhance personalization, considering factors like location and activity (Almonte et al., 2021). Graph-Based Systems rely on networks or graphs to represent user–item connections, providing accurate recommendations and addressing cold start and data sparsity challenges. Hybrid Systems combine multiple methods for improved accuracy but come with increased processing times and operational costs (Guo et al., 2020; Lakshmi & Bhavani, 2021; Wang et al., 2021).

In conclusion, various RS cater to different needs, each with its advantages and limitations. Hybrid Systems offer a comprehensive approach, combining the strengths of different methods to enhance recommendation accuracy (Lakshmi & Bhavani, 2021; Shani & Gunawardana, 2011).

4.2.2.3 Performance Evaluation Metrics of Recommendation Systems

In addition to the measures outlined in Section 4.2.2.1., which continue to pose challenges that RS must overcome, according to Lakshmi and Bhavani (2021), as well as Shani and Gunawardana (2011), there are numerous metrics to evaluate the performance of RS depending on the type and objective of the evaluation RS. We will only highlight the following metrics that assess the accuracy of predicted ratings, as they are most relevant to the type of RS implemented in this chapter. Thus, the evaluation metrics for the accuracy of predicted ratings will be as follows:

- Root of Mean Squared Error (RMSE):

$$\text{RMSE} = \left[\frac{\sum_{i=1}^{n} \left| \hat{y}_i - y_i \right|^2}{n} \right]^{\frac{1}{2}} = \left[n^{-1} \sum_{i=1}^{n} \left| e_i \right|^2 \right]^{\frac{1}{2}}, \tag{4.1}$$

$$\text{where } |e_i| = |\hat{y}_i - y_i|$$

RMSE is a widely used accuracy measure in recommendation systems. It assesses the precision of predicted ratings by comparing them to actual ratings. RMSE is also employed to compare prediction errors across different models. The goal is to minimize RMSE, with a value closer to 0 indicating better model performance. To calculate RMSE for RS, a test dataset (*T*) with real user ratings is defined. Each real rating is temporarily omitted, and the RS predicts its value. This process is repeated for all cases in the test set, allowing the calculation of RMSE or Mean Absolute Error (MAE) (Shani & Gunawardana, 2011; Willmott & Matsuura, 2005).

Therefore, considering a test set *T* of user–item pairs (*u, i*) for which the real ratings (r_{ui}) are unknown, the RS will predict the values of these ratings ($\hat{r}_{ui}$) and calculate the RMSE using the following formula:

$$\text{RMSE} = \left[\frac{1}{T} * \sum_{(u,i)\in T} \left(\hat{r}_{ui} - r_{ui}\right)^2 \right]^{\frac{1}{2}} \tag{4.2}$$

- Mean Absolute Error (MAE):

MAE calculates the arithmetic mean of absolute errors, where these errors are obtained by taking the difference between the predicted value ($\hat{y}_i$) and the expected value (y_i) for instances representing the same phenomena (Fürnkranz et al., 2011; Willmott & Matsuura, 2005).

$$\text{MAE} = \frac{\sum_{i=1}^{n} |\hat{y}_i - y_i|}{n} = \left[n^{-1} \sum_{i=1}^{n} |e_i| \right], \tag{4.3}$$

$$\text{MLN} = \left(V_M,\ E_M, V, L\right)$$

In fact, the process required to calculate the difference between the predicted ratings and the actual ratings is the same as indicated for RMSE. However, the formula for calculating MAE to evaluate the studied RS is as follows:

$$\text{MAE} = \left[\frac{1}{T} * \sum_{(u,i)\in T} |\hat{r}_{ui} - r_{ui}| \right] \tag{4.4}$$

It is an alternative to the RMSE measure; however, RMSE is a disproportionate measure that tends to penalize large errors. In other words, when there is a significant variation between the rating predicted by the SR and the actual rating, the RMSE measure tends to highlight these types of errors more than MAE (Shani & Gunawardana, 2011).

4.2.3 Network Science

Considering all the previously mentioned models, we will focus on Graph-Based RS, as these represent interactions between users and different items through networks. They consider possible relationships between these elements to generate recommendations, which have proven to be relevant and efficient in the context of EC (Guo et al., 2020; Lakshmi & Bhavani, 2021; Wang et al., 2021).

Graph theory defines graphs or networks as consisting of nodes (vertices) connected by edges. These connections can be represented mathematically using adjacency or incidence matrices. Graphs can have weighted or binary connections and may be direct or indirect. Attributes can be associated with both nodes and links, allowing for analysis at different scales (Sporns, 2018; Yang et al., 2021).

4.2.3.1 Single-Layer Networks

A single-layer network, or monolayer network, often referred to as a one-mode network is characterized by having only one layer. These are the "standard" networks, which can be represented by a graph $G = (V, E)$, where V (vertices) is a set of nodes and E (edges) corresponds to the connections established between different pairs of nodes, i.e., $E \subseteq V \times V$ (Barabási & Pósfai, 2016; Fogelman-Soulié et al., 2020).

To represent a network or a graph, a common approach is to use an adjacency matrix (A), where $A_{ij} = 1$ if node i is connected to node j and $A_{ij} = 0$ otherwise (Barabási & Pósfai, 2016; Fogelman-Soulié et al., 2020).

One of the advantages of using single-layer networks is the abundance of studies and works in this area. As a result, there are numerous algorithms that have been researched and implemented in these networks, including centrality measures, network motifs, clique detection, community detection, among others (Santra et al., 2021). However, the limitation of representing only a single layer makes single-layer networks less effective in capturing networks and graphs with multiple characteristics and different entities. To address this issue, it becomes relevant to turn to MLNs.

4.2.3.2 Multilayer Networks

Multilayer networks (MLN) are defined as a network interconnected by multiple layers, which is composed not only of nodes and edges but also of various types of layers. These networks can also include additional features, such as aspects, which correspond to the different types of interactions or relationships among the nodes. (Böttcher & Porter, 2021; Kivela et al., 2014).

Kivela et al. (2014) define a MLN as a quadruple (Figure 4.2):[i]

$$\text{MLN} = \left(V_M,\ E_M, V, L\right)$$

where

1. $V_M \subseteq V \times L$ specifies the entity, i.e., the node–layer combination in which a node is present in the corresponding layer. Additionally, there are nodes that are absent in certain layers, and therefore, it becomes useful to indicate whether nodes are present in a certain layer, which is done through V_M;
2. E_M represents the set of connections in a MLN, and it is represented by the set of pairs of possible combinations between nodes and elementary layers, i.e., $E_M \subseteq V_M \times V_M$;

3. V refers to a specific node;
4. $L = L_1 \times \ldots \times L_d$ corresponds to the set of layers in the MLN. L_l indicates which layer of the MLN we are referring to or the layer in which the node V is present.

Several authors assert that it is relevant to distinguish between the different types of connections that can exist in a MLN (Kivela et al., 2014; Nasiri et al., 2021). These connections can include pairwise connections between all possible combinations of nodes and layers. In other words, a particular node $\boldsymbol{u}$ in layer $\boldsymbol{\alpha}$ can be connected to any node $\boldsymbol{v}$ in any layer $\boldsymbol{\beta}$. That is, there can be connections between different layers, known as interlayer connections. Additionally, there can be connections between a node $\boldsymbol{x}$ and another node $\boldsymbol{y}$ within the same layer, in this case called intralayer connections (Figure 4.2).

There are also coupling edges ($E_{\tilde{C}} \subseteq E_C$), which correspond to connections where two nodes (u and v) represent the same entity in different layers. With this, it becomes possible to define the different types of graphs according to the type of connections they exhibit: intralayer graphs such as $G_A = (V_M, E_A)$; interlayer graphs such as $G_C = (V_M, E_C)$; and coupling graphs such as $G_{\tilde{C}} = (V_M, E_{\tilde{C}})$.

Another important point to mention in a MLN is the concept of aspects that correspond to groups of layers of different types (Kivela et al., 2014; McGee et al., 2019).

MLNs can be classified as either direct or indirect, depending on the type of connections they represent (Kivela et al., 2014). Unlike single-layer networks, implementing MLNs poses challenges in developing algorithms for execution, performance

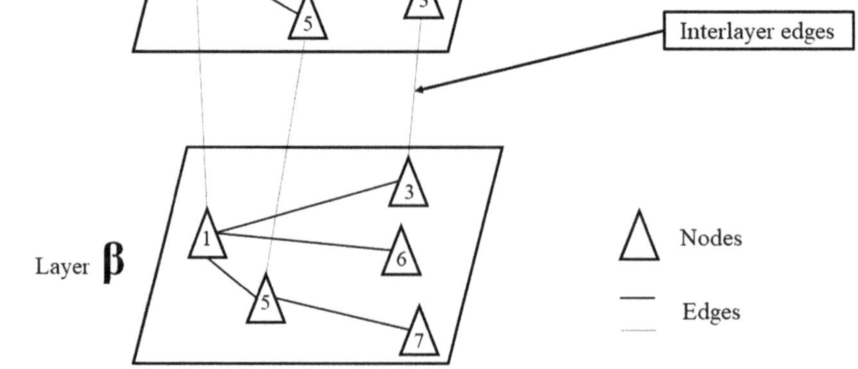

FIGURE 4.2 Example of a multilayer network.

(Source: Own elaboration.)

analysis, and understanding results. Balancing information from node attributes and preserving network structure without sacrificing either is a key challenge, often requiring significant computational capacity (Santra et al., 2021).

4.2.3.3 Bipartite Networks

Bipartite networks (BNs), also referred to as two-mode networks, are heterogeneous networks characterized by nodes belonging to two distinct entities or families, where connections occur exclusively between nodes from different entities; in other words, two nodes from the same entity are never connected (Arthur, 2020; Daud et al., 2020; Fogelman-Soulié et al., 2020; Lv et al., 2023).

In the context of EC and RS, we always work with users and items, which essentially translates into two types of nodes from different families or entities. One family of nodes are related to users or clients and another family of nodes represent items or products. Examining more closely, when a transaction occurs, i.e., when a user purchases an item, this transaction can be visualized or represented in a graph, as a connection is observed between the user's node and the corresponding node of the acquired product (Lakshmi & Bhavani, 2021).

Considering nodes from two different entities, I and O, respectively, we can represent the graph of a bipartite network as $G = (I, O, E)$, where E represents the set of relationships between the different families of nodes. In other words, considering node $i \in I$ and node $o \in O$, if node i is connected to node o, then $(i, o) \in E$. Thus, the representation of the adjacency matrix (A) of a BN has dimensions $m \times n$, where m corresponds to the number of nodes in set I and n is the number of nodes in set O. Here, $A_{io} = 1$ if node $i \in I$ is connected to node $o \in O$, and $A_{io} = 0$ otherwise, meaning the nodes are not connected (Fogelman-Soulié et al., 2020).

One of the major challenges of BNs arises in their analysis, as many algorithms and existing metrics are optimized for one-mode networks. For this reason, there are two ways to analyse a BN (Arthur, 2020):

- the first and most commonly used approach involves projecting a two-mode network onto a one-mode network;
- the second approach involves modifying and adapting algorithms initially suitable for single-layer networks to make them effective for bipartite network analysis. However, this option is not very common because algorithms designed for one-mode networks often cannot be directly applied to BNs.

Based on the previous definition of two-mode network graphs: $G = (I, O, E)$, it can be projected onto two one-mode networks: $G_I = (I, E_I)$ and $G_O = (O, E_O)$ from the sets of nodes I and O separately (Fogelman-Soulié et al., 2020).

4.2.3.4 Link Prediction

Link prediction in network analysis infers future connections based on existing network links and topological properties of nodes, enhancing predictions for missing or likely-to-form relationships (Daud et al., 2020; Rahman et al., 2017).

In our methodology, we leverage network structures to anticipate future connections and infer missing links. For instance, as illustrated in Figure 4.3, even when

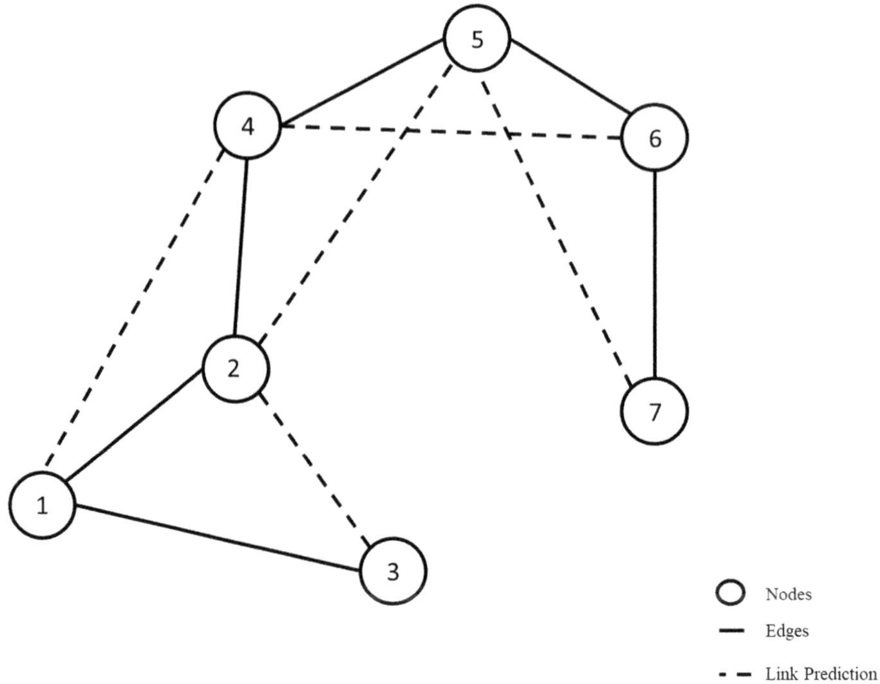

FIGURE 4.3 Example of link prediction.

(Source: Own elaboration.)

nodes 2 and 5 are not directly connected, their shared link to node 4 suggests a potential future connection between them. Additionally, we address missing links that may emerge or vanish within a network during a specific time interval (t to $t + 1$) using link prediction techniques.

Both content-based measures, which utilize node attributes, and graph-based measures, which leverage the topological structure of networks, are employed for LP to determine the probability of a future connection between two unconnected nodes (Daud et al., 2020). Daud et al. (2020) divide LP approaches into four different categories, classified as:

- Similarity: Local indices, quasi-local indices, or global indices;
- Probabilistic: Probability tensor factorization model, Markov model, or link label modelling;
- Algorithmic: Meta-heuristic, matrix factorization, or machine learning;
- Hybrid.

Hao et al. (2020) state that there are two major types of LP: transductive link prediction (T-LP) and inductive link prediction (I-LP), as seen in Figure 4.4. The T-LP focuses on predicting new connections between nodes that already exist in the graph but were previously unconnected.

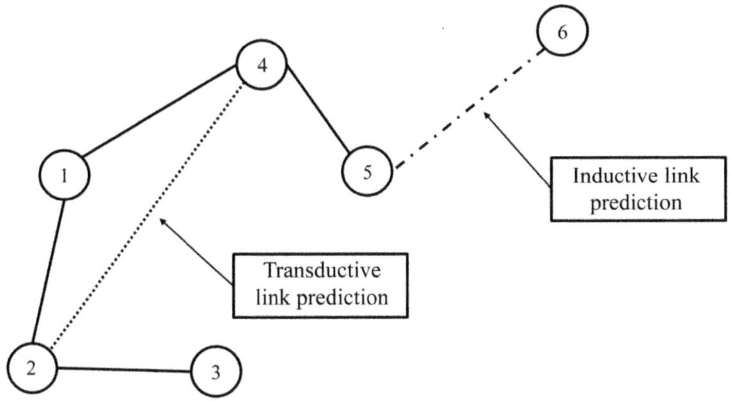

FIGURE 4.4 Transductive vs inductive link prediction.

(Source: Figure adapted from Hao et al., 2020.)

Researchers often consider both content-based and graph-based measures for T-LP (Hao et al., 2020). These measures leverage information from both the network structure and node attributes to enhance prediction accuracy. The goal of I-LP is to predict connections involving a new node that is not part of the existing graph. Since this new node lacks connections within the graph structure, graph-based measures are not applicable. Instead, I-LP relies solely on node attributes to determine similarity between nodes based on their intrinsic characteristics. Implementing I-LP techniques is crucial for addressing the cold start problem, a significant challenge in RS (Hao et al., 2020; Yang et al., 2021).

Concerning MLNs, neglecting the relationship between various layers and their influence on the LP process between two nodes in a particular layer can lead to a significant loss of information and accuracy in predicting new connections. Therefore, when addressing LP in a MLN, it is crucial to consider not only intralayer information but also interlayer information. This approach ensures better performance and increased accuracy in predicting new connections (Luo et al., 2021; Najari et al., 2019).

Existing LP methods optimized for one-mode or homogeneous networks are often ineffective in BNs due to their distinct node types (Zhou et al., 2022). To overcome this, the solution involves one of two approaches: projecting the BN onto a one-mode network or adapting existing algorithms.

In recommendation systems, temporal measures are crucial due to evolving user preferences. Factors like evaluation time and acquisition duration impact link prediction accuracy (Lakshmi & Bhavani, 2021).

4.3 METHODOLOGY

4.3.1 Data

In the development of this process, a hypothetical database was used to build the network. This database consists in 246 ratings from 15 different users (Norris, Jéssica, Lewis, Nancy, Hamilton Amanda, Emily, Mark, Alonso, George, Lando, Jennifer,

John, Lucy, and Sainz) and 96 distinct items, categorized into three item types: A, B, and C. Each row in this database corresponds to a rating (ranging from 1 to 5) assigned by a user to a respective item belonging to a category (the same user may give different ratings to the same product if they purchase that item more than once over time, and their taste or experience with the product changes). The structure of the dataset is composed of: Category I User I Product I Rating.

For data processing and implementation of the proposed method, the Cross Industry Standard Process for Data Mining (CRISP-DM) model was used. This model divides the process into six phases, as shown in Shearer (2000): understanding the domain under study, understanding the data, data preparation, modelling, evaluation, and implementation. To know more, one can consult Ramoa (2022).

4.3.2 Visualization of Recommendations in Two One-Mode Multilayer Networks Connected by Bipartite Links

As we saw during the literature review, in recommendations, users and items can be represented by nodes in a BN, and the links in this network correspond to interactions or transactions between items and users. In this case, LP can be applied to predict a potential connection between users and items, improving the quality of information by recovering missing links or establishing new connections (Stanhope et al., 2019).

However, if we separate the connections between these entities by different categories of items or products, then the initial bipartite network can also be represented as a MLN, where each layer corresponds respectively to a category of items or products, as seen in Figure 4.5.

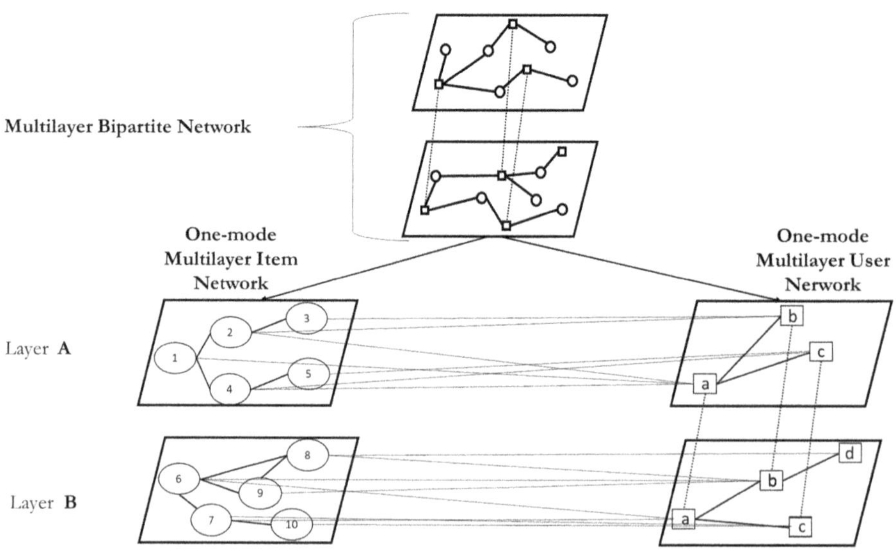

FIGURE 4.5 Representation of users and items by category through a multilayer bipartite network.

(Source: Own elaboration.)

If from this representation we decide to separate the BN (two-mode networks) into two one-mode networks, respectively for items and users, then the network visualization can be represented according to the scheme presented in Figure 4.6. It is worth noting that in the example shown in Figure 4.6, we restricted the network to only two categories, but we can use as many categories as we want.

As mentioned in category A of Figure 4.5, user A purchased products 1, 2, and 4. If we consider a one-mode network of items, these items will be connected to each other. This process is repeated successively for all users until we have the one-mode network of items represented in category A of Figure 4.6. The same occurs in the implementation of the one-mode user network.

When superficially analysing the one-mode networks presented in Figure 4.6, they may seem to have a more confusing visualization/understanding. However, upon closer inspection, it becomes apparent that this representation translates into information gains and greater clarity.

For example, in this representation, we still have connections between the two types of nodes, i.e., between items and users. These connections are established between each of the MLN, resulting in a kind of inter-network connections. It is through these connections that the final LP will ideally be obtained, allowing the prediction of new purchases for consumers and the implementation of cross-selling and up-selling strategies.

Observing Figure 4.6, it is noted that the interlayer connections between categories A and B in the multilayer one-mode network of users are fixed. This is because,

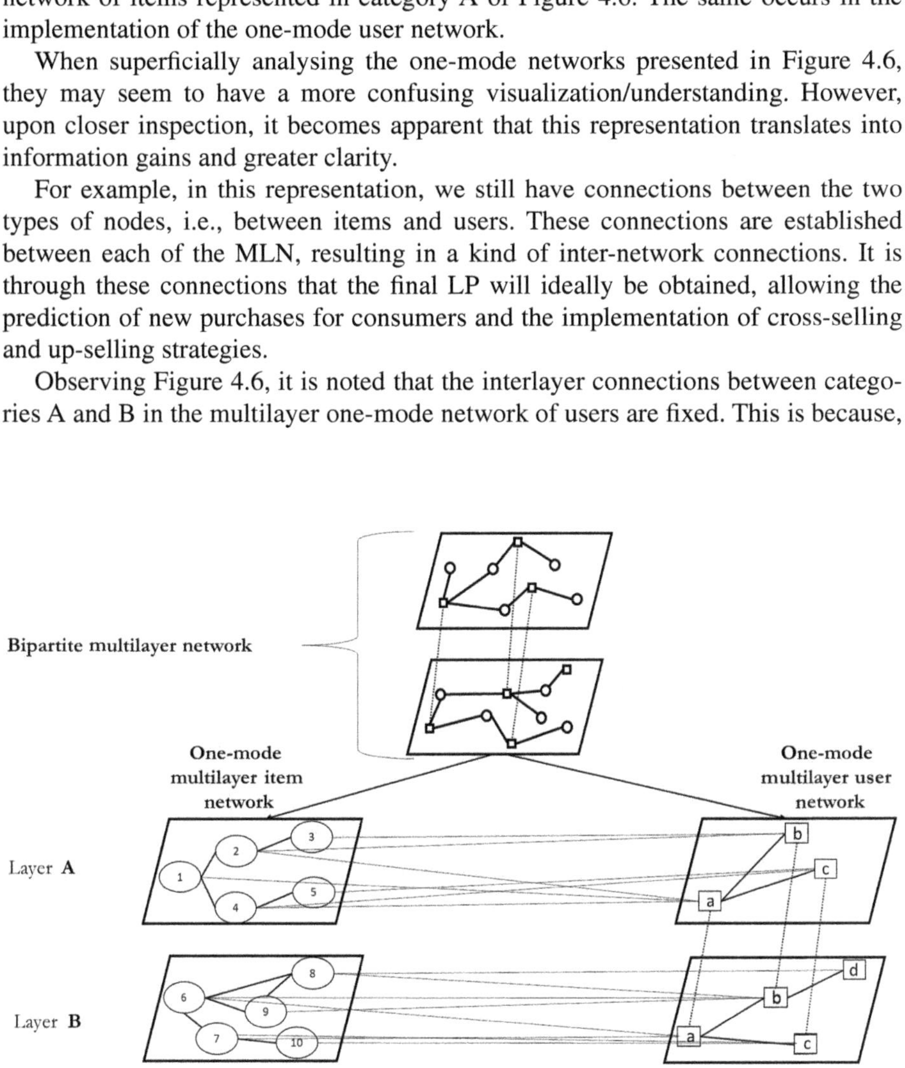

FIGURE 4.6 From multilayer bipartite network to multilayer one-mode networks – example with two categories.

(Source: Own elaboration.)

as some users engage in transactions across different categories, they end up being repeatedly included in those layers of the network, forming these connections.

Regarding the LP process to anticipate new consumer/user purchases (corresponding to the inter-networks link prediction in Figure 4.7), the ideal algorithm should include a mechanism to align all information derived from both attributes and the network structure, as well as from the other predictions established when performing LP in each of the multilayer one-mode networks. This alignment mechanism could involve a loss function, as applied by authors Hao et al. (2020) and Yang et al. (2021) in MLNs. It is worth noting that in Figure 4.7, dashed lines indicate examples of predicted connections in the LP process, while the remaining connections represent existing connections.

I-LP, being one of the LP processes especially useful in RS, will be one of the most relevant LP processes to consider when implementing an algorithm to predict new consumer purchases.

However, it is not sufficient to only include information from attributes for the LP process. Hence, the use of T-LP mechanisms will be another crucial aspect to consider when predicting new purchases for users. An example of this LP can be visualized in Figure 4.7, in category A of the multilayer one-mode user network, where there is a prediction of a connection between users b and c, who already existed in the graph.

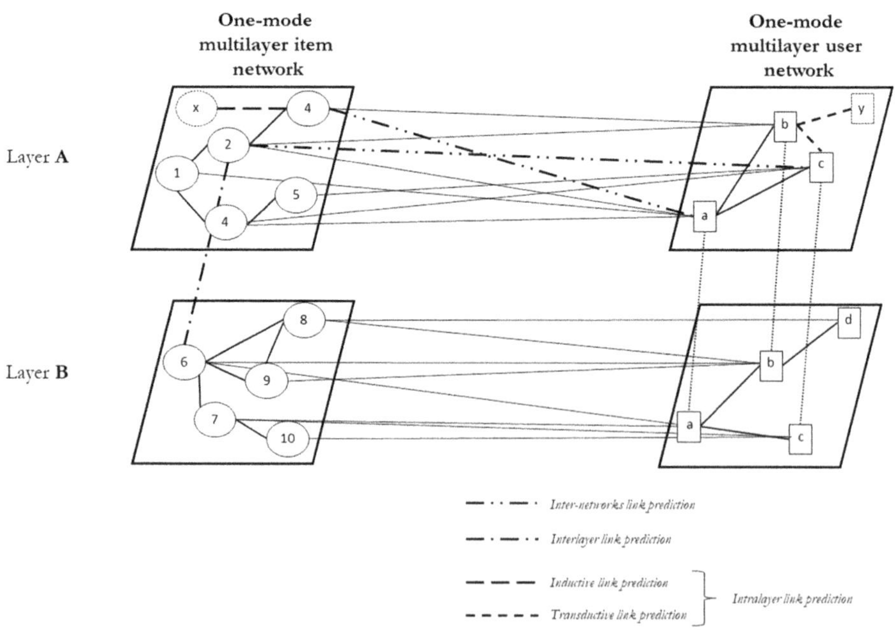

FIGURE 4.7 Link prediction in and between two multilayer one-mode networks connected to each other by bipartite links.

(Source: Own elaboration.)

Focusing now only on the projected multilayer (one-mode) item network, predicting new connections in this network will enable the implementation of new cross-selling and up-selling strategies. It is important to always consider the users' profiles and what they are looking for, as in this example, the connections will tend to be direct links between node 2 and node 6, where the reciprocal may not occur.

In the context of predicting connections between multilayer one-mode networks, an ideal algorithm should: incorporate LP (Link Prediction) principles to improve accuracy; integrate both attribute and structural information, taking into account connection types (interlayer or intralayer); and balance implementation complexity, execution time, and computational capacity to achieve optimal precision and performance.

4.3.3 Implementation of a Hybrid Recommendation System

In this chapter, we implement a hybrid method of CF that is simultaneously user-based and item-based, as it utilizes the k-means and weighted slope one algorithms, respectively. The choice for the algorithm of the slope one family was due to the fact that it does not require great computational capacity and allows to overcome one of the great problems of RS, data sparsity.

The biggest drawbacks of the slope one algorithms stem from the fact that they are item-based; that is, they focus only on the items, leaving aside the users, and the similarities and resemblances between their profiles. With the aim of overcoming this situation and increasing the performance of the implementation of the slope one algorithms, Song and Wu (2020) begin by implementing an improved clustering method, called k-means++ (a method where the distance measure is cosine similarity). Subsequently, based on these groups, they implement the weighted slope one algorithm. Yannam et al. (2022) follow the same method but use traditional k-means. In turn, Li et al. (2011) use k-nearest-neighbourhood methods to combine the similarities between users before the implementation of the slope one algorithms. In reality, this method could be seen as a hybrid RS, since the final recommendation obtained directly depends on the simultaneous implementation of these two techniques, respectively, user-based and item-based.

Based on the literature review and part of Section 4.3.2, one of the first adaptations is to think of this problem as a B-MLN and implement this method always based on network thinking. The starting point will be a transaction database that includes all item categories, where we have a table with users, items, and their respective ratings or scores assigned by users to the respective items purchased. According to Lakshmi and Bhavani (2021), Collaborative Filtering Systems techniques use a rating matrix as input, where columns correspond to items, rows to users, and cells to ratings. From our database a bipartite adjacency matrix will then be created. It should be noted that this rating is the equivalent in a graph to the weight of the connections between nodes, a weight that is an attribute of the connections. The final matrix will correspond to a bipartite adjacency matrix composed in the rows only by the nodes corresponding to the users and in the columns only by the nodes corresponding to the items, or vice versa.

The values that make up the matrices correspond to the ratings assigned to the items by the users, which generally comprise values between 1 and 5 (Song & Wu, 2020).

When a user has not made any transaction with a certain item, or has not assigned it any score, then this case will be represented in the matrix by the rating with value 0. When we have the final bipartite adjacency matrix, we can then move on to the implementation of the RS.

Since not all users are active at the same time on EC sites, it does not make sense for the RS to use part of its implementation time and computational capacity to predict recommendations for users who are offline. For this reason, one of the first steps to do in the algorithm is to define the target user or users (u) who will correspond to the users who are with the active session, browsing the EC site, which will be feasible to implement through the identification of their ID. Thus, the final objective will be to predict the recommendations for the items that the target users (u) did not purchase or did not assign a score, as demonstrated by the final formulas of the flowchart presented in Figure 4.8, respectively, $\boldsymbol{P}^{WS_1}_{(u)_j}$ and $\boldsymbol{P}^{WP}_{(u)_j}$.

For this, we start by implementing the k-means algorithm which aims to group users by clusters and determine: if the target user u belongs to a cluster that contains similar users, that is, if it is a cluster with a length greater than 1; or if it does not have users similar to it, that is, if the length of the cluster is equal to 1, which indicates that the user u is the only constituent element of that cluster. To effectively execute the k-means, it is necessary to previously find the ideal number of clusters, which can be achieved through the elbow method (Umargono et al., 2020).

The distance measure used in k-means was the Euclidean, as this is the most used when it comes to numerical attributes, such as in this case are the ratings (Umargono et al., 2020). If we notice, with the use of the clustering algorithm (k-means) what we are doing can be interpreted as a task of LP in the one-mode network of users in Figure 4.7. Given that the final bipartite adjacency matrix will contain all categories, then this LP process will produce both new intralayer and interlayer connections.

As we can verify in the flowchart of Figure 4.8, from the moment we have the clusters formed it is possible to determine which group u belongs to. On one hand, if u belongs to a cluster that has a length of 1, it means that the group consists only of itself, which is due to the fact that there are no other users in the matrix that are similar to u. To determine which items to recommend for this same user, we follow the implementation of the original weighted slope one algorithm created by Lemire and Maclachlan (2005), starting by finding the average deviation between items j and items i ($\boldsymbol{dev}_{j,i}(\mathbf{1})$), for all users of the matrix (indicated by s) who assigned a score to both items.

However, in order to save computational time, the items j will only correspond to the items that the target users did not previously assign any score, which makes the calculations of the algorithm more selective and decreases its execution time.

That is, we now have both target users indicated by u and their target items given by j. When we have the $dev_{j,i}(1)$ calculated for all desired items j, we can then predict the ratings that the target users would possibly assign to these items j through the calculation of the weighted slope one indicated by $\boldsymbol{P}^{WS_1}_{(u)_j}$.

On the other hand, if the length of the cluster to which u belongs is greater than 1, it means that this group has more than 1 element, that is, the target user u has similar users, being able to establish connections between them. In this way, considering the

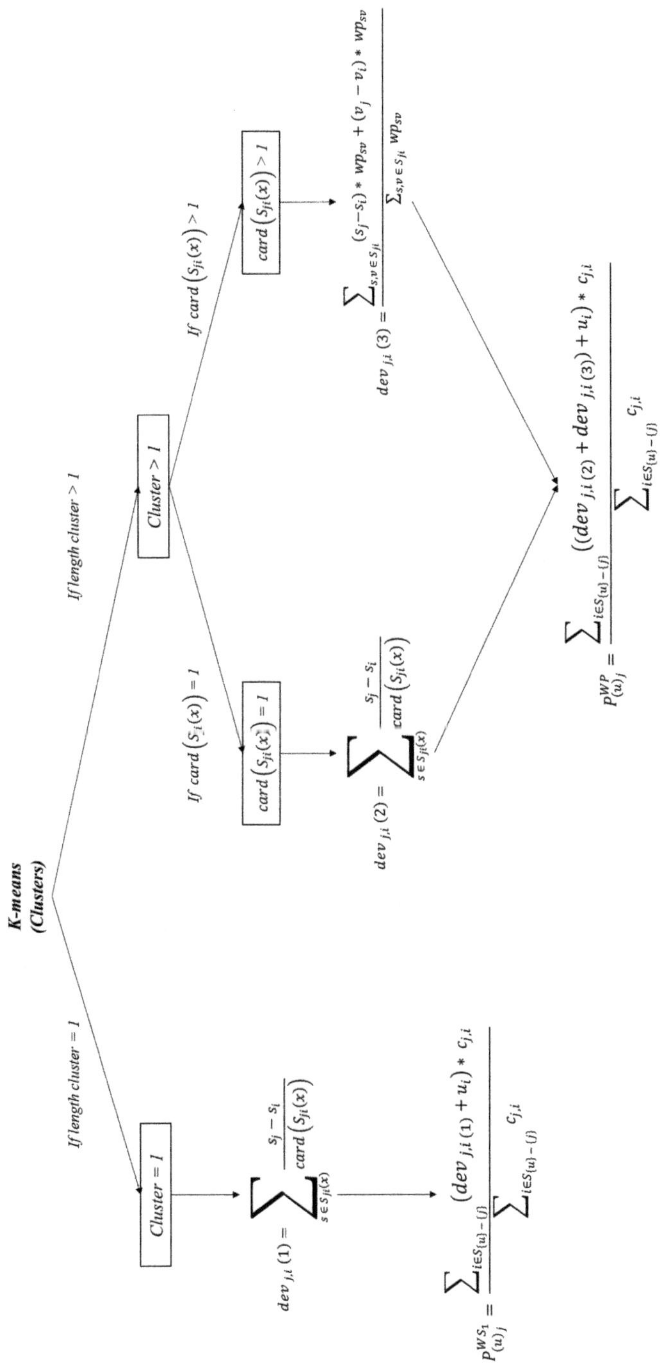

FIGURE 4.8 Flowchart of the implemented recommendation system.

(Source: Own elaboration.)

same target user u and the items j to which the user did not assign any score, to calculate the $dev_{j,i}$, we have to take into account two situations:

1. If only one similar user has rated both items j and i, we use the traditional deviation formula ($\boldsymbol{dev_{j,i}}$**(2)**).
2. If multiple similar users (e.g., s and v) have rated items "j" and "i", the average deviation ($\boldsymbol{dev_{j,i}}$**(3)**) presented in Figure 4.8 is calculated. This average deviation is adapted from Song and Wu (2020) and considers variations in ratings for both items across all users in the cluster, multiplied by the standard deviation factor (*wp_sv*), which represents the stability of the ratings assigned between users s and v.

In this situation, the recommendations are predicted according to the formula we adapted from Song and Wu (2020) named Weighted Slope One Algorithm of Improved Fusion Score Preference $\left(\boldsymbol{P}^{\boldsymbol{WP}}_{(u)_j}\right)$.

After determining predicted scores for items unrated by target users, we proceed to generate final recommendations. Items with the highest positive scores are recommended and displayed to target users during their active session on the EC site.

With the purpose of determining the target items for recommendation, we implement, as done by Song and Wu (2020) and Yannam et al. (2022), a Top-N recommendation or item list, aiming to filter predicted ratings by the RS specifically for items with the highest predicted values for the target user. The average rating scale typically ranges from 1 to 5, with 3 as the midpoint. Consequently, recommended items should have scores above this average, aligning with the user's preferences. To operationalize this, we divide the Top-N recommendation process into two stages based on the average ratings assigned by target users: if the average of previously assigned ratings for purchased products is greater than or equal to 3.5, all items with predicted values at or above this average are potential recommendations; if the average of target users' ratings falls below 3.5, all items with predicted scores greater than or equal to 3 should be recommended to these users.

We integrate the final part of the RS implemented in this study with the methodology proposed in Figure 4.7 of Section 4.3.2. After using k-means to predict intralayer connections within the one-mode user network, establishing relationships between users, we implement weighted slope one based on the information obtained from k-means. Finally, we predict new inter-network connections, as depicted in Figure 4.7. These emerging inter-network connections guide the recommendations for target users, resulting in an inter-networks LP process. Further details about the RS implemented can be seen in Ramoa (2022).

4.3.4 Software

For the implementation of this work, an Intel Core i7 tenth-generation computer with 16 GB of RAM was used. The fictitious database was created in Microsoft Excel 2016. Data modelling and model implementation were performed using RStudio software with version RStudio 2022.02.3+492 and the R program with version 4.2.1 (R Core Team, 2022). Various R packages were implemented for this purpose: **igraph** – used to transform the dataset into a network and create the bipartite adjacency matrix (Csardi & Nepusz, 2006); **tidyverse** – utilized for data manipulation and modelling (Wickham et al., 2019); **factoextra & NbClust** – employed in the implementation of the k-means algorithm to determine the optimal number of clusters (Kassambara & Mundt, 2020; Malika Charrad et al., 2014); **metrics** – used to obtain measurement functions for evaluating the studied RS (Hamner & Frasco, 2018).

4.4 DATA ANALYSIS AND RESULTS

The process of data preparation and modelling involves creating a table from an original dataset, averaging scores assigned by users, and constructing a bipartite adjacency matrix for RS. Once we obtain the final bipartite adjacency matrix, we can implement the proposed RS, starting with the first algorithm, k-means, which groups users. To determine the ideal number of clusters, we use the elbow rule method. As seen in Figure 4.9, the ideal number of clusters lies between four and six. We choose to divide it into six clusters to have greater heterogeneity than if we chose four or five.

After the implementation of k-means according to the ideal number of clusters, six groups were formed with the following composition of users: **Group 1**: Sainz; **Group 2**: Hamilton, Jennifer, Jéssica, and Lando; **Group 3**: Norris; Nancy; **Group 4**: John, Lewis, Lucy, and Mark; **Group 5**: Alonso, Amanda, and George; **Group 6**: Emily.

Through the formation of these clusters, we are, in fact, establishing LP between different users in the one-mode user network shown in Figure 4.7. For example, considering Group 2, with the creation of these clusters, Hamilton, Jennifer, Jéssica, and Lando would now be connected to each other through an intralayer connection.

Once the groups are established, six target users were chosen for whom the second algorithm of the proposed RS, in this case, the weighted slope one, will predict the items to recommend. In other words, it will predict new purchases for these consumers based on the predicted ratings. The selection of these target users took

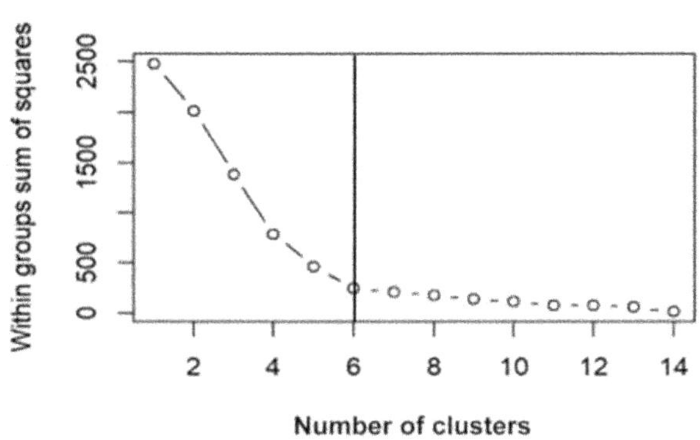

FIGURE 4.9 Elbow method (ideal number of clusters).

(Source: Own elaboration.)

into account the formed groups, meaning one member from each cluster was chosen, enabling an understanding of how the different aspects of the RS proposed in this chapter. Thus, the target users are: Sainz, Hamilton, Norris, Lewis, Amanda, and Emily.

After the application of the RS, another bipartite adjacency matrix was obtained, which includes both the previously assigned ratings by users and the predicted ratings for the target users. From this matrix, Top-N recommendations were made – based on the actual average ratings assigned by the target users. It is important to highlight that the ratings obtained essentially stem from a link prediction process and correspond to the new inter-network connections shown in Figure 4.7, where the weight of these connections corresponds to the rating value. Furthermore, these highlighted items also represent the prediction of new purchases for consumers, in this case, for the target users described above. When recommended, these items are the most likely to be purchased by the target users, according to the proposed RS.

4.4.1 Evaluation of the Recommendation System

For evaluation, we employ the RMSE and MAE measures. These metrics align well with the type of recommendation system under study, as previously discussed in Section 4.2.2.3. Through an evaluation process similar to the "leave-one-out cross-validation" method, we begin by eliminating or hiding (setting to zero) each real rating value (r_{ui}), turning it into an item for which the user in question would not have assigned any score, i.e., into an unknown rating (Lemire & Maclachlan, 2005). In the second step, we try, through our model, to predict what the predicted value would be for this now unknown rating $\left(\hat{r}_{ui}\right)$. This process was repeated for each user and for each item for which there would be a user–item pair with a real assigned rating, which corresponds to our test set (T), in this case consisting of the total of 246 real ratings in the database.

The predicted values obtained for the previously hidden ratings were then incorporated into a new matrix with the values of $\hat{r}_{ui}$, which through the difference with the values contained in the original r_{ui} matrix allowed us to calculate the MAE and RMSE measures, according to what is stated in equations (4.2) and (4.4), respectively. The MAE value obtained was 1.87, which indicates that the difference between the predicted rating value through our model and the actual rating value is on average approximately 1.87. The RMSE value obtained was 1.43, and thus, the fact that the RMSE is smaller than the MAE suggests that most of the model's prediction errors are not very significant. It would be important to understand whether the value obtained by the MAE is high or low, considering the limits or the range of the ratings. Despite the system's predictions not being bound by the usual 1 to 5 range, we can still recommend items based on them. If the predicted rating is negative or below 3, those items should not be recommended. If it's above 3, regardless of the specific value, those are the items to recommend. The RMSE doesn't necessarily exceed the MAE, suggesting consistent predictions, which may indicate that there are not many discrepant variations between the obtained values and the actual values. While these metrics are typically used for comparing different recommendation systems, they still provide valuable insights here.

4.5 FINAL CONCLUSIONS, LIMITATIONS, AND FUTURE CONSIDERATIONS

EC significantly impacts global economies and consumer experiences. The implementation of advanced RS can be integrated as part of digital transformation to enhance the overall customer experience, aligning with modern consumer expectations. Our research explored projections of bipartite one-mode MLNs onto interconnected networks, offering an alternative RS approach. We successfully implemented a hybrid RS using k-means and weighted slope one algorithms, addressing data sparsity and predicting new consumer purchases.[ii] However, limitations include using a small fictitious database.

Future work should explore real databases and incorporate implicit and explicit data for deeper insights. Leveraging node attributes and I-LP techniques can mitigate the cold start problem in RS. Through this analysis of bipartite one-mode MLN, we gain a multidimensional view of customers, enabling cross-selling strategies and overcoming RS challenges, such as cold start.

In conclusion, optimizing supply chain processes using data-driven insights contributes to the efficiency and responsiveness of a digitally transformed business. It can also lead to the creation of innovative business models and novel business opportunities and revenue streams. Implementing real-time decision support systems, informed by RS and network analysis, is a part of the transformative journey towards agile and responsive business operations.

NOTES

i The way to define a multilayer network is not fixed, varying according to the author. However, in this chapter we ended up adopting this definition because we consider it to be the one that describes this type of network in more detail.

ii The R programming code used to develop the model presented here is available on request from the authors.

REFERENCES

Alamdari, P. M., Navimipour, N. J., Hosseinzadeh, M., Safaei, A. A., & Darwesh, A. (2020). A systematic study on the recommender systems in the E-commerce [article]. *IEEE Access*, *8*, 115694–115716. https://doi.org/10.1109/access.2020.3002803

Almonte, L., Guerra, E., Cantador, I., & de Lara, J. (2021). Recommender systems in model-driven engineering A systematic mapping review [Article; Early Access]. *Software and Systems Modeling*, 32. https://doi.org/10.1007/s10270-021-00905-x

Alrumiah, S. S., & Hadwan, M. (2021). Implementing big data analytics in E-commerce: Vendor and customer view [article]. *IEEE Access*, *9*, 37281–37286. https://doi.org/10.1109/access.2021.3063615

Arthur, R. (2020). Modularity and projection of bipartite networks. *Physica A – Statistical Mechanics and Its Applications*, *549*, Article 124341. https://doi.org/10.1016/j.physa.2020.124341

Ballestar, M. T. (2021). Editorial: Segmenting the future of E-commerce, one step at a time. *Journal of Theoretical and Applied Electronic Commerce Research*, *16*(2), I–III. https://doi.org/10.4067/s0718-18762021000200101

Barabási, A., & Pósfai, M. (2016). *Network Science*. Cambridge University Press, Cambridge.

Behera, R. K., Gunasekaran, A., Gupta, S., Kamboj, S., & Bala, P. K. (2020). Personalized digital marketing recommender engine. *Journal of Retailing and Consumer Services*, *53*, 101799.

Blattberg, R. C., Kim, B.-D., & Neslin, S. A. (2008). Cross-selling and up-selling. In *Customer Equity: Building and Managing Relationships as Valuable Assets*, (pp. 515–547). Springer, New York. https://doi.org/10.1007/978-0-387-72579-6_21

Böttcher, L., & Porter, M. A. (2021). Classical and quantum random-walk centrality measures in multilayer networks. *SIAM Journal on Applied Mathematics*, *81*(6), 2704–2724. https://doi.org/10.1137/20m1385998

Cerqueira, T. G., Ramalho, F., & Marinho, L. B. (2016). *A Content-Based Approach for Recommending UML Sequence Diagrams*. Knowledge Systems Institute, SEKE.

Charrad, Malika, Ghazzali, Nadia, Boiteau, Veronique, & Niknafs, Azam (2014). *NbClust: An R Package for Determining the Relevant Number of Clusters in a Data Set*. In http://www.jstatsoft.org/v61/i06/

Coppola, D. (2021). *E-commerce Worldwide – Statistics & Facts*. Statista. Retrieved 25-10-2021 from https://www.statista.com/topics/871/online-shopping/#dossier-chapter1

Csardi, G., & Nepusz, T. (2006). The igraph software package for complex network research. *InterJournal*, 2006, 169.

Das, S., Patra, B. K., & Kumar, J. (2021). Weighted slope one with threshold filtering. In *Advances in Data Science and Intelligent Data Communication Technologies* (pp. 143–150). Springer, Singapore. https://doi.org/10.1007/978-981-15-8061-1_12

Daud, N. N., Ab Hamid, S. H., Saadoon, M., Sahran, F., & Anuar, N. B. (2020). Applications of link prediction in social networks: A review. *Journal of Network and Computer Applications*, *166*, 102716. https://doi.org/10.1016/j.jnca.2020.102716

Fayyaz, Z., Ebrahimian, M., Nawara, D., Ibrahim, A., & Kashef, R. (2020). Recommendation systems: Algorithms, challenges, metrics, and business opportunities. *Applied Sciences*, *10*(21), 7748.

Fogelman-Soulié, F., Mei, L., Zhang, J., Li, Y., Ge, W., Li, Y., & Ye, Q. (2020). Recommender systems and attributed networks. *Advances in Data Science: Symbolic, Complex and Network Data*, *4*, 139–167.

Fürnkranz, J., Chan, P. K., Craw, S., Sammut, C., Uther, W., Ratnaparkhi, A., Jin, X., Han, J., Yang, Y., Morik, K., Dorigo, M., Birattari, M., Stützle, T., Brazdil, P., Vilalta, R., Giraud-Carrier, C., Soares, C., Rissanen, J., Baxter, R. A., … De Raedt, L. (2011). Mean absolute error. In *Encyclopedia of Machine Learning and Data Mining* (pp. 652–652). Springer, US. https://doi.org/10.1007/978-0-387-30164-8_525

Guo, Q., Zhuang, F., Qin, C., Zhu, H., Xie, X., Xiong, H., & He, Q. (2020). A survey on knowledge graph-based recommender systems. *IEEE Transactions on Knowledge and Data Engineering*, *1–1*. https://doi.org/10.1109/tkde.2020.3028705

Hamner, B., & Frasco, M. (2018). *Metrics: Evaluation Metrics for Machine Learning*. In (Version R package version 0.1.4) https://CRAN.R-project.org/package=Metrics

Hao, Y., Cao, X., Fang, Y., Xie, X., & Wang, S. (2020). *Inductive Link Prediction for Nodes Having Only Attribute Information* Association for Computing Machinery (ACM), New York.

He, G. H. (2021). Enterprise E-commerce marketing system based on big data methods of maintaining social relations in the process of E-commerce environmental commodity. *Journal of Organizational and End User Computing*, *33*(6), Article 16. https://doi.org/10.4018/JOEUC.20211101.oa16

Išoraitė, M., & Miniotienė, N. (2018). *Electronic Commerce: Theory and Practice*, IGI Global, Hershey, PA.

Kassambara, A., & Mundt, F. (2020). *factoextra: Extract and Visualize the Results of Multivariate Data Analyses*. In https://CRAN.R-project.org/package=factoextra

Kivela, M., Arenas, A., Barthelemy, M., Gleeson, J. P., Moreno, Y., & Porter, M. A. (2014). Multilayer networks. *Journal of Complex Networks*, *2*(3), 203–271. https://doi.org/10.1093/comnet/cnu016

Kosasih, E. (2023). *Neurosymbolic Reasoning for Link Prediction in Supply Chain Knowledge Graphs*, Springer, Cham, Switzerland.

Lakshmi, T. J., & Bhavani, S. D. (2021). Link prediction approach to recommender systems. *arXiv preprint* https://doi.org/arxiv:2102.09185

Lemire, D., & Maclachlan, A. (2005). *Slope One Predictors for Online Rating-Based Collaborative Filtering, ACM (Association for Computing Machinery), New York, NY.*

Lesage, L., Deaconu, M., Lejay, A., Meira, J. A., Nichil, G., & State, R. (2020). A recommendation system for car insurance. *European Actuarial Journal*, *10*(2), 377–398. https://doi.org/10.1007/s13385-020-00236-z

Li, J., Sun, L., & Wang, J. (2011). A slope one collaborative filtering recommendation algorithm using uncertain neighbors optimizing. *International Conference on Web-Age Information Management.*

Luo, H., Li, L., Zhang, Y., Fang, S., & Chen, X. (2021). Link prediction in multiplex networks using a novel multiple-attribute decision-making approach. *Knowledge-Based Systems*, *219*, 106904. https://doi.org/10.1016/j.knosys.2021.106904

Lv, H., Zhang, B., Li, T., & Hu, S. (2023). Construction and analysis of multi-relationship bipartite network model. *Complex & Intelligent Systems*, *9*(5), 5851–5863. https://doi.org/10.1007/s40747-023-01038-y

Mathews, J., Mehta, P., Suryamukhi, M., & Babu, S. (2021). Link prediction techniques to handle tax evasion. *Proceedings of the 3rd ACM India Joint International Conference on Data Science & Management of Data (8th ACM IKDD CODS & 26th COMAD).*

McGee, F., Ghoniem, M., Melançon, G., Otjacques, B., & Pinaud, B. (2019). The state of the art in multilayer network visualization. *Computer Graphics Forum*, *38*(6), 125–149. https://doi.org/10.1111/cgf.13610

Najari, S., Salehi, M., Ranjbar, V., & Jalili, M. (2019). Link prediction in multiplex networks based on interlayer similarity. *Physica A: Statistical Mechanics and its Applications*, *536*, 120978. https://doi.org/10.1016/j.physa.2019.04.214

Nasiri, E., Berahmand, K., & Li, Y. (2021). A new link prediction in multiplex networks using topologically biased random walks. *Chaos, Solitons & Fractals*, *151*, 111230. https://doi.org/10.1016/j.chaos.2021.111230

R Core Team. (2022). R: A language and environment for statistical computing. In *R Foundation for Statistical Computing*. Vienna, Austria. https://www.R-project.org/

Rahman, M. S., Dey, L. R., Halder, S., Uddin, M. A., Islam, M., & IEEE. (2017). Link prediction by correlation on social network. *[2017 20th International Conference of Computer and Information Technology (ICCIT)]. 20th International Conference of Computer and Information Technology (ICCIT)*, Univ Asia Pacific, Dept Comp Sci & Engn, Dhaka, Bangladesh.

Ramoa, L. (2022). *Sistemas de Recomendação em E-Commerce: Link Prediction em Redes Multilayer Bipartidas.* School of Economics and Management of the University of Porto. https://hdl.handle.net/10216/146599

Ricci, F., Rokach, L., & Shapira, B. (2015). Recommender systems: Introduction and challenges. In (pp. 1–34). Springer, US. https://doi.org/10.1007/978-1-4899-7637-6_1

Santra, A., Komar, K., Bhowmick, S., & Chakravarthy, S. (2021). From base data to knowledge discovery – A life cycle approach – Using multilayer networks. https://arxiv.org/abs/2105.11410

Sarwar, B., Karypis, G., Konstan, J., & Reidl, J. (2001). *Item-based Collaborative Filtering Recommendation Algorithms*, Association for Computing Machinery (ACM), New York, NY.

Shani, G., & Gunawardana, A. (2011). Evaluating recommendation systems. In *Evaluating Recommendation Systems*, (pp. 257–297). Springer, US. https://doi.org/10.1007/978-0-387-85820-3_8

Shearer, C. (2000). The CRISP-DM model: The new blueprint for data mining. *Journal of Data Warehousing*, *5*(4), 13–22.

Smith, B., & Linden, G. (2017). Two decades of recommender systems at Amazon.com. *IEEE Internet Computing*, *21*(3), 12–18.

Song, Y. T., & Wu, S. (2020). Slope one recommendation algorithm based on user clustering and scoring preferences. *Procedia Computer Science*, *166*, 539–545.

Sporns, O. (2018). Graph theory methods: applications in brain networks. *Dialogues in Clinical Neuroscience*, *20*(2), 111–121. https://doi.org/10.31887/dcns.2018.20.2/osporns

Stanhope, A., Sha, H., Barman, D., Al Hasan, M., & Mohler, G. (2019). Group link prediction. *IEEE International Conference on Big Data [2019 ieee international conference on big data (big data)]. IEEE International Conference on Big Data (Big Data)*. Los Angeles, CA.

Turban, E., Outland, J., King, D., Lee, J. K., Liang, T.-P., & Turban, D. C. (2018). *Electronic Commerce 2018: A Managerial and Social Networks Perspective*. Springer, Cham, Switzerland.

Umargono, E., Suseno, J. E., & Vincensius Gunawan, S. K. (2020). *K-Means Clustering Optimization Using the Elbow Method and Early Centroid Determination Based on Mean and Median Formula*, IEEE, Piscataway, NJ.

Vadwala, A. Y., & Vadwala, M. M. S. (2017). E-Commerce: Merits and demerits a review paper. *International Journal of Trend in Scientific Research and Development*, *1*(4), 117–120.

Vikatos, P., Gryllos, P., & Makris, C. (2020). Marketing campaign targeting using bridge extraction in multiplex social network. *Artificial Intelligence Review*, *53*(1), 703–724.

Wang, S., Hu, L., Wang, Y., He, X., Sheng, Q. Z., Orgun, M. A., Cao, L., Ricci, F., & Yu, P. S. (2021). Graph learning based recommender systems: A review. arXiv preprint arXiv:2105.06339.

Wickham, H., Averick, M., Bryan, J., Chang, W., McGowan, L., François, R., Grolemund, G., Hayes, A., Henry, L., Hester, J., Kuhn, M., Pedersen, T., Miller, E., Bache, S., Müller, K., Ooms, J., Robinson, D., Seidel, D., Spinu, V., … Yutani, H. (2019). Welcome to the tidyverse. In https://doi.org/10.21105/joss.01686

Willmott, C., & Matsuura, K. (2005). Advantages of the mean absolute error (MAE) over the root mean square error (RMSE) in assessing average model performance. *Climate Research*, *30*, 79–82. https://doi.org/10.3354/cr030079

Yang, S., Hu, B., Zhang, Z., Sun, W., Wang, Y., Zhou, J., Shan, H., Cao, Y., Ye, B., Fang, Y., & Yu, Q. (2021). Inductive link prediction with interactive structure learning on attributed graph. In *Advances in Data Science and Intelligent Data Communication Technologies* (pp. 383–398). Springer International Publishing. https://doi.org/10.1007/978-3-030-86520-7_24

Yang, Y., Wang, F., Zhu, E., Jiang, F., & Yao, W. (2024). Social behavior analysis in exclusive enterprise social networks by FastHAND. *ACM Transactions on Knowledge Discovery from Data*, 18(2), 1–22.

Yannam, V. R., Kumar, J., Babu, K. S., & Patra, B. K. (2022). Enhancing the accuracy of group recommendation using slope one. *The Journal of Supercomputing*. https://doi.org/10.1007/s11227-022-04664-4

Zhou, C., Zhang, J., Gao, K., Li, Q., Hu, D., & Sheng, V. S. (2022). Bipartite network embedding with symmetric neighborhood convolution. *Expert Systems with Applications*, *198*, 116757.

5 Digital Transformation and Business Processes

An Experience-Based Perspective

João Pedro Ladeiras
Deloitte Portugal, Porto, Portugal

Carolina Machado
University of Minho, Braga, Portugal

5.1 INTRODUCTION

In an era where digital experiences permeate both our daily lives and organizational landscapes, the synergy between technology and business processes has become more crucial than ever. The advent of digital transformation has revolutionized how businesses operate, compelling them to adapt to evolving technological landscapes to stay competitive and relevant. This paradigm shift not only impacts operational efficiency but also reshapes customer interactions and employee workflows.

This chapter delves into the intricate relationship between digital transformation and business processes, exploring the evolution of technology as a driving force behind this transformation. From the emergence of low-code platforms to the rise of end-to-end automation, key players in the digital arena are reshaping traditional workflows and operational paradigms.

One of the focal points of this exploration is the profound impact of low-code and end-to-end automation technologies on cost reduction and operational efficiency. By streamlining processes and minimizing manual intervention, organizations can achieve unprecedented levels of productivity and agility.

Furthermore, this chapter elucidates the shift toward employee-centric and customer-centric processes, emphasizing the importance of aligning technological advancements with the needs and expectations of both internal stakeholders and external customers.

However, the journey toward digital transformation is not without its challenges and risks. Through the lens of end-to-end automation and low-code platforms, this text navigates the complexities and potential pitfalls organizations may encounter along this transformative journey.

In addition to dissecting the current landscape, in this chapter we also look to offer strategic insights into digital transformation, particularly focusing on end-to-end

DOI: 10.1201/9781003461821-5

integration and automation. By leveraging robust strategies, organizations can navigate the complexities of digital transformation and unlock new avenues for growth and innovation.

Finally, this text peers into the future, exploring emerging trends in end-to-end integration and automation. From AI-driven processes to blockchain-enabled workflows, the future promises unprecedented possibilities for organizations willing to embrace digital transformation.

In short, while digital transformation is reshaping the way we live, work, and interact, company processes and systems face a significant challenge in keeping pace with the seamless and user-friendly processes and applications we've come to expect in our daily lives. Therefore, join us on this journey as we unravel the intricate tapestry of digital transformation and business processes, illuminating the path toward a more agile, efficient, and customer-centric future.

5.2 UNDERSTANDING DIGITAL TRANSFORMATION

Digital transformation can be defined as the integration of digital technology into all areas of a business, fundamentally changing how you operate and deliver value to customers. It's also a cultural change that requires organizations to continually challenge the status quo, experiment, and get comfortable with failure (Zhu, Ge and Wang, 2021; Zaoui and Souissi, 2020; Reis, Amorim, Melão and Matos, 2018).

Conceptually, digital transformation is not solely about technology. Instead, it's about reshaping organizations to be more agile, innovative, and customer- and employee-focused (Li, Gao, Han, Gupta, Alhalabi and Almakdi, 2023). These changes should enable the organization to respond to varying market demands rapidly, focus on internal processes for efficiency, and create a seamless, integrated customer and employee experience.

This concept has been around since the beginning of digital computers, but it gained significant attention with the rise of the internet and e-commerce in the late 1990s and early 2000s. Businesses began to realize the potential of using digital tools not only for operational efficiency but also for strategic advantage.

In the early stages, digital transformation was often equated with digitization—the process of converting analog information into digital form. However, as technology evolved, so did the understanding of digital transformation. It expanded to include digitalization, which involves using digital technology and digitized data to impact how work gets done, transform how customers and companies engage and interact, and create new (digital) revenue streams.

5.2.1 Understanding Business Processes

Traditional business processes are often characterized by manual operations, hierarchical decision-making structures, and siloed departments. While these processes may have been effective in the past, they come with several limitations in the digital era (Harmon, 2003):

Inefficiency: Manual processes are time-consuming and prone to errors. They can also lead to bottlenecks, especially when decisions have to pass through multiple levels of management.

Lack of Agility: Traditional business processes are often rigid and inflexible, making it difficult for businesses to adapt to changes in the market.

Lack of Integration between Systems: Traditional business processes often involve multiple, disconnected systems. This lack of integration can lead to data silos, inefficiencies, and errors.

Poor Employee Experience: If core systems are difficult to use, it can lead to a poor employee experience, which can impact productivity and morale.

Poor Customer Experience: Siloed departments can lead to a disjointed customer experience, as different teams may not have access to the same information or may not communicate effectively with each other.

5.3 DIGITAL EXPERIENCE ON DAILY LIVES AND IN ORGANIZATIONS

In our daily lives, we're witnessing a rapid technology transformation. Mobile applications have become an integral part of our routines, enabling us to perform a multitude of tasks at the touch of a button. From ordering food and hailing a taxi to managing our finances and staying connected with friends, technology has seamlessly integrated into our personal lives.

However, this level of digital transformation often doesn't translate as smoothly into the organizational context. Many organizations struggle to keep pace with the rapid advancements in technology. This disparity between the digital experiences in our personal lives and those in our work lives is often referred to as the "Digital Divide".

Many organizations face challenges when trying to keep the same pace on internal processes as for customer experience due to (Heavin and Power, 2018; Kraus, Durst, Ferreira, Veiga, Kailer and Weinmann, 2022):

Legacy Systems: Many organizations still rely on outdated systems and processes that are not compatible with newer technologies. Upgrading these systems can be costly and time-consuming.

Resistance to Change: People are often resistant to change, especially when it comes to technology. Employees may be comfortable with existing processes and reluctant to learn new systems.

Security Concerns: Organizations have to deal with sensitive data, making security a top priority. While new technologies can offer better functionality, they may also present new security risks.

Regulatory Compliance: Certain industries are heavily regulated. Organizations in these industries must ensure that any new technology or process complies with all relevant laws and regulations, which can slow down the adoption of new technologies.

Employee Experience: Often, organizations concentrate their efforts on enhancing revenue streams and, as a result, customer experiences. However, they tend to overlook the importance of refining their back-office processes and internal systems. This oversight can lead to the creation of isolated, non-integrated processes that lack an end-to-end perspective, subsequently resulting in increased operational costs.

5.4 THE EVOLUTION OF TECHNOLOGY IN DRIVING DIGITAL TRANSFORMATION AND BUSINESS PROCESSES

The transformative journey of digital technology within the context of business processes has witnessed a remarkable evolution, shifting from heavy code reliance to the advent of low-code technologies with end-to-end automation capabilities.

In the earlier stages of digital transformation, organizations heavily relied on traditional coding practices to develop and implement technological solutions. This approach, while effective, often posed challenges related to time, cost, and agility. Developing and maintaining complex lines of code required specialized expertise and extended timelines, limiting the speed at which businesses could adapt to dynamic market demands. Automation capabilities were often compartmentalized and focused on specific tasks or processes.

The evolution toward low-code technologies marked a paradigm shift in the development landscape. Low-code platforms empower organizations to create applications with minimal manual coding, streamlining the development process and significantly reducing time-to-market (Rokis and Kirikova, 2022). These platforms often feature visual interfaces, pre-built components, and drag-and-drop functionality, democratizing the application development process and enabling a broader range of stakeholders to contribute to digital initiatives. This allows the creation of sophisticated workflows, integration with diverse systems, and orchestration of complex processes—all with reduced manual coding requirements. The democratization of automation empowers a broader spectrum of users within the organization to contribute to and benefit from automated workflows, enabling businesses to automate end-to-end business processes. This not only improves efficiency and productivity but also reduces the risk of errors and enhances customer experiences.

These new technologies will empower organizations to accelerate digital transformation. This acceleration is particularly evident when considering business process optimization. Low-code platforms empower individuals with domain expertise but limited coding skills, often referred to as citizen developers, to actively participate in the development and automation of business processes. This democratization of technology enables a more inclusive approach to innovation, tapping into the collective intelligence of diverse teams. It also enhances the flexibility and adaptability of business processes. Organizations can easily modify and optimize workflows based on evolving requirements without extensive coding efforts. This adaptability aligns with the dynamic nature of modern business environments.

5.4.1 Key Players in Low-Code and End-to-End Automation

Several key players stand out in the arena of low-code development platforms and end-to-end automation solutions. These platforms are revolutionizing the way businesses operate, enabling them to rapidly develop and deploy digital solutions with minimal coding effort.

Some of the relevant platforms of the market include the following (Martinez and Pfister, 2023; Alsaadi, Radain, Alzahrani, Alshammari, Alahmadi and Fakieh, 2021):

OutSystems has established itself as a prominent player in the low-code application development space. Offering visual development tools, pre-built templates, and artificial intelligence (AI)-driven suggestions, OutSystems accelerates the development lifecycle, allowing organizations to bring applications to market at unprecedented speeds. With a focus on both web and mobile applications, OutSystems addresses the need for versatile solutions. The platform's strength lies in its ability to support complex integrations, making it a valuable asset for organizations looking to bridge the gap between legacy systems and modern digital capabilities.

Microsoft Power Platform stands as a comprehensive suite, encompassing Power Apps, Power Automate, Power BI, and Power Virtual Agents. This integrated offering provides organizations with a versatile toolkit for low-code application development, workflow automation, business analytics, and chatbot creation. Leveraging the synergy within the Microsoft ecosystem, Power Platform empowers businesses to seamlessly connect disparate systems, automate repetitive tasks, and derive actionable insights from data. The platform's user-friendly interfaces and pre-built templates democratize the development process, enabling a broader range of users to actively contribute to digital initiatives.

Pega positions itself as a relevant player in the context of intelligent automation, providing a unified platform for low-code application development, business process management (BPM), and customer relationship management (CRM). Pega's strength lies in its ability to infuse AI and machine learning (ML) into business processes, creating a dynamic environment where automation is not merely rule-based but intelligent and adaptive.

Appian provides a low-code automation platform that accelerates the creation of high-impact business applications. It combines the power of low-code development with advanced automation capabilities, enabling businesses to rapidly transform their operations. Appian positions itself at the intersection of low-code development and BPM. The platform's emphasis on collaborative design makes it a preferred choice for those seeking to integrate process automation with application development. Appian's approach empowers organizations to create sophisticated workflow solutions that align with their unique business requirements.

Mendix, a Siemens business, offers a comprehensive set of tools for the entire application life cycle, from ideation and development to deployment and operation. Mendix positions itself as a comprehensive low-code application development platform designed to accelerate the creation of web and mobile applications. With a focus on empowering both professional developers and business users, Mendix aims to bridge the collaboration gap, allowing diverse stakeholders to actively contribute to the development process.

ServiceNow provides a suite of cloud-based services to automate business processes. Its platform includes a low-code development environment that allows businesses to build applications tailored to their specific needs. ServiceNow has a platform that goes beyond conventional information technology service management solutions, offering a unified approach to workflows across various business functions, focusing particularly on enterprise service management.

UiPath, a frontrunner in the robotic process automation (RPA) space, brings a distinct focus on end-to-end automation. Specializing in automating repetitive and rule-based tasks, UiPath's low-code platform empowers organizations to design and deploy automation workflows across diverse business processes. The platform's user-friendly interface and strong community support contribute to its popularity. UiPath's capabilities extend beyond traditional RPA, providing organizations with the tools needed for a holistic approach to automation, aligning with the broader goals of digital transformation.

5.5 THE IMPACT OF LOW-CODE AND END-TO-END TECHNOLOGIES ON COST REDUCTION AND OPERATIONAL EFFICIENCY

Low-code and end-to-end automation platforms contribute significantly to cost reduction (Sanchis, García-Perales, Fraile and Poler, 2020). By simplifying the application development process, these platforms reduce the need for extensive coding, which in turn reduces the time and resources required for application development. This leads to significant cost savings, as businesses can develop and deploy applications faster and with fewer resources. Moreover, these platforms often come with built-in features for scalability and maintenance, further reducing the total cost of ownership.

In terms of operational efficiency, low-code and end-to-end automation platforms enable businesses to automate repetitive and time-consuming tasks, freeing up employees to focus on more strategic and value-adding activities. This not only improves productivity but also enhances the quality of work, as the risk of human error is minimized. Furthermore, by enabling real-time data collection and analysis, these platforms provide businesses with actionable insights that can be used to optimize operations and improve decision-making.

These platforms foster agility and innovation, as they enable businesses to quickly adapt to changing market conditions and customer needs, which is crucial in today's fast-paced digital economy. By empowering non-technical users to participate in the application development process, these platforms also democratize innovation, leading to more diverse and creative solutions.

These technologies are able to provide a relevant impact on process transformation through (Sanchis, García-Perales, Fraile and Poler, 2020):

Accelerated Development Cycles: These platforms empower organizations to build applications swiftly with minimal manual coding. The accelerated development cycles translate to reduced time-to-market.

Democratized Development: By lowering the barriers to entry in application development, low-code platforms democratize the development process. This not only fosters collaboration between business and IT but also expands the pool of contributors, reducing the dependency on specialized developers and potentially lowering development costs.

Resource Optimization: Traditional development often requires a dedicated team of developers, architects, and testers. Low-code platforms streamline these roles, allowing developers to focus on more complex tasks, while business users can actively contribute to the development process. This resource optimization contributes to cost reduction.

Holistic Process Optimization and Seamless Integration across Functions: End-to-end automation platforms enable organizations to optimize their business processes holistically. These technologies eliminate the compartmentalization of processes by establishing unified workflows that seamlessly traverse departmental boundaries. Silos, once barriers to smooth operations, are dismantled, fostering a cohesive operational environment. The reach of these platforms extends beyond organizational limits to integrate vendors, customers, and employees into the operational ecosystem.

Security and Compliance: Low-code and end-to-end automation platforms often come with built-in security features and compliance standards. This is particularly important for businesses operating in regulated industries, such as finance and healthcare.

Risk Mitigation through Standardization: Standardizing processes through automation mitigates the risk of errors and inconsistencies. This not only enhances operational efficiency but also safeguards organizations against potential financial and reputational risks associated with manual errors.

Advanced Analytics Integration: End-to-end automation platforms often integrate advanced analytics capabilities. This allows organizations to derive actionable insights from automated processes, facilitating data-driven decision-making across various functions, from supply chain optimization to customer relationship management.

Real-Time Monitoring and Reporting: The automation of workflows enables real-time monitoring of key performance indicators. This continuous visibility into operational processes empowers organizations to identify bottlenecks, optimize resource allocation, and make informed decisions promptly.

Besides process transformation, compliance, and efficiency gains, by using these solutions, organizations will be able to:

Reduce Development Costs: With low-code platforms, organizations can achieve significant cost reductions in development efforts. As mentioned before, the reduced need for extensive coding expertise contributes to overall cost savings.

Achieve Operational Cost Efficiencies: End-to-end automation reduces operational costs by minimizing manual intervention in routine tasks. Efficiency gains in areas such as customer service, human resources (HR) processes, and supply chain management lead to tangible cost efficiencies.

Experience Faster Time-to-Value: The combination of low-code development and end-to-end automation expedites the delivery of applications and the implementation of automated workflows. This faster time-to-value accelerates the realization of cost reductions and operational efficiencies.

Scalability: These platforms are designed to scale with the needs of the business. As the organization grows, the platforms can handle increased demand without a significant increase in operational costs. This scalability is crucial for businesses that are experiencing rapid growth or have fluctuating demand.

5.6 EMPLOYEE-CENTRIC AND CUSTOMER-CENTRIC PROCESSES

In the era of digital transformation, businesses are shifting their focus from product-centric to people-centric, placing customers and employees at the heart of their strategies (Shah, Rust, Parasuraman, Staelin and Day, 2006). This paradigm shift is driven by the integration of digital technology into all areas of a business, fundamentally changing how they operate and deliver value to customers.

The integration of customer-centric and employee-centric processes goes beyond mere coexistence—it signifies a strategic alignment that permeates every aspect of organizational functioning. For example, a unified digital ecosystem forms the bedrock where customer interactions and employee experiences seamlessly intersect. Technologies that drive customer engagement are mirrored in those that empower employees, creating a cohesive and interconnected environment. This allows us to break down silos and make processes converge through cross-functional collaboration, fostering a holistic understanding of organizational dynamics from both client and employee perspectives.

A good example of a process where these platforms can break down silos and unify the end-to-end process vision, with a focus on employees and customers, is the Order-to-Cash (O2C) process (Kaaria and Shamsuzzoha, 2024; Serrala, 2019). This process considers the entire lifecycle of a customer order, from the initial order placement to the final receipt of payment. It is a key component of the broader revenue cycle management within an organization. The O2C process involves various stages, including order entry, order fulfillment, shipping, invoicing, and, ultimately, the collection of payment.

Here's a breakdown of the key stages in the Order-to-Cash process (Kaaria and Shamsuzzoha, 2024; Doxey, 2019):

Order Entry: The process begins when a customer places an order for goods or services. This involves capturing essential details such as product specifications, quantities, pricing, and delivery preferences.

Order Fulfillment: Once an order is received, the organization works to fulfill it by sourcing or producing the requested goods or services. This stage involves inventory management, production planning, and coordination across various departments to ensure timely delivery.

Shipping and Invoicing: After the order is fulfilled, the products are shipped to the customer. Simultaneously, an invoice is generated detailing the products or services provided, along with the corresponding costs. This document serves as a formal request for payment.

Payment Collection: The final stage involves collecting payment from the customer. This may include various methods such as electronic payments, checks, or other agreed-upon terms. Efficient payment collection is crucial for maintaining healthy cash flow and sustaining the organization's financial health.

In a customer-centric context, these technologies facilitate a seamless and personalized experience throughout the O2C lifecycle. Automation ensures real-time order processing, reducing lead times and enhancing order fulfillment efficiency. Low-code platforms enable organizations to quickly adapt their systems to changing customer demands, providing flexibility in tailoring order processes. Customer interactions, from order placement to payment, become smoother, more responsive, personalized, and based on a unified system.

Moreover, end-to-end automation optimizes the invoicing and payment collection stages. Automated invoicing systems generate accurate and timely invoices, minimizing errors and delays. Integration with payment gateways streamlines the collection process, providing customers with various payment options for a convenient and efficient experience.

On the employee-centric front, low-code platforms empower non-technical staff to actively participate in shaping and optimizing the O2C process. Employees can easily modify workflows, design user interfaces, and implement changes without extensive coding knowledge. This democratization of the development process not only accelerates the implementation of improvements but also fosters a culture of continuous improvement and innovation.

Additionally, end-to-end automation reduces the burden on employees by automating routine tasks, allowing them to focus on more strategic and value-added activities (Zhang, Fan, He and Zhang, 2021). This not only enhances job satisfaction but also contributes to skill development as employees engage with the technology to optimize and innovate within the O2C process.

The unified data ecosystem created by these technologies ensures that both customers and employees benefit from a holistic understanding of the O2C process. Insights derived from automated analytics enable organizations to refine strategies, address pain points, and align the O2C process with evolving expectations.

By analyzing the O2C example, it is clear that through seamless portals and streamlined workflows, these platforms enhance the overall customer and employee experience, ensuring personalized interactions. The unified platform facilitates a holistic view of the end-to-end process, breaking down silos and fostering collaboration across departments. It transforms data into actionable insights, empowering organizations to focus on analyzing new process data. This data-driven approach becomes essential for continuous improvement initiatives.

5.7 CHALLENGES AND RISKS THROUGH THE DIGITAL TRANSFORMATION JOURNEY USING END-TO-END AUTOMATION AND LOW-CODE PLATFORMS

Transforming business processes with end-to-end automation and low-code platforms undoubtedly brings significant advantages, but it is crucial to recognize the challenges and risks during the transformation journey to ensure a successful transformation (Martinez and Pfister, 2023; Sanchis, García-Perales, Fraile and Poler, 2020).

Implementing end-to-end automation requires a deep understanding of business processes and often involves integrating multiple systems and technologies. This can be complex and time-consuming, requiring significant resources and expertise. A key challenge is that these platforms cannot inherently fix an inefficient process. Businesses must ensure their processes are efficiently designed before automation, or the automation will only magnify the inefficiencies.

These platforms are not a cure-all for flawed processes. They offer efficiency gains and streamlined workflows, but they cannot fix a poorly designed process. If the underlying business processes are inefficient, automation and end-to-end integration may amplify existing issues. Organizations must invest time in thoughtful process design and optimization before leveraging automation and end-to-end integration to ensure they enhance overall efficiency.

There is a risk of overreliance on these platforms as a universal solution. They may not be suitable for every business process, especially complex or highly specialized ones, which may require tailored solutions. Organizations need to assess their unique process requirements thoroughly before embarking on an automation journey.

The rapid evolution of technology introduces the risk of platform obsolescence. Organizations need to select platforms that align with their long-term strategic goals and can adapt to emerging technological trends. Failure to anticipate future needs and advancements may result in costly retooling or replacement of platforms, impacting financial resources and operational continuity.

Integration challenges pose another significant risk. Harmonizing automation and end-to-end integration with existing systems and legacy infrastructure can be complex. Compatibility issues, data migration challenges, and the need for extensive customization can disrupt business operations. A thorough integration strategy, including phased implementations and continuous monitoring, is essential to mitigate these risks effectively.

Security concerns are a constant challenge. While automation offers efficiency gains, it also increases the attack surface. Inadequate security measures can expose organizations to cyber threats, data breaches, and unauthorized access. Robust cybersecurity protocols, regular audits, and ongoing vigilance are imperative to safeguard sensitive information and maintain stakeholder trust.

5.8 DIGITAL TRANSFORMATION STRATEGIES FOR END-TO-END INTEGRATION/AUTOMATION

Digital transformation is a strategic imperative for businesses in the modern era. It involves the integration of digital technology into all areas of a business, fundamentally changing how you operate and deliver value to customers. End-to-end automation and low-code platforms are key enablers of this transformation, offering significant benefits such as increased efficiency, reduced costs, and improved agility.

Successful digital transformation requires a well-thought-out strategy (Matt, Hess and Benlian, 2015). This strategy should be based on a clear understanding of the business's goals, the needs of its customers, and the capabilities of its technology. It should also take into account the unique challenges and opportunities presented by end-to-end automation and low-code platforms.

One effective strategy is to start small and scale up. This involves identifying a specific process or area of the business that could benefit from automation or low-code development, implementing the technology, and then scaling up based on the lessons learned and the results achieved. This approach allows businesses to realize immediate benefits while also building the skills and capabilities needed for larger-scale transformations.

Another key strategy is to foster a culture of innovation and continuous learning (Leal-Rodríguez, Sanchís-Pedregosa, Moreno-Moreno and Leal-Millán, 2023; Fischer, Lundin, Lindberg, 2020). This involves encouraging employees to experiment with new technologies, learn from their successes and failures, and continuously improve their skills and processes. This culture can help businesses to adapt to the rapid pace of technological change and to harness the full potential of end-to-end automation and low-code platforms.

In terms of frameworks for planning, executing, and sustaining digital initiatives, there are several that businesses can consider.

Agile Methodology: The Agile methodology is well-suited for digital transformations, offering a flexible and iterative approach. Breaking down projects into small, manageable tasks allows for quick adaptation to changing requirements, ensuring that the organization remains responsive to evolving needs (Chaudhari and Joshi, 2021).

Digital Maturity Models: Leveraging digital maturity models provides a structured framework for assessing an organization's current digital capabilities and charting a roadmap for improvement. These models help organizations understand where they stand in their digital transformation journey and identify areas for enhancement (Alsufyani and Gill, 2021).

Change Management Frameworks: Successful digital transformation is as much about managing people and culture as it is about adopting new technologies. Change management frameworks provide a structured approach to anticipate, manage, and address resistance, ensuring a smoother transition for employees and stakeholders (Phillips and Klein, 2023).

Continuous Improvement Models: A culture of continuous improvement is fundamental to sustained digital transformation. Models such as the Plan–Do–Check–Act (PDCA) cycle or Six Sigma provide methodologies for ongoing assessment, refinement, and optimization of digital processes (Lameijer, Boer, Anthony and Does, 2023).

Technology Adoption Frameworks: Selecting and adopting new technologies, including end-to-end automation and low-code platforms, require careful consideration. Frameworks that assess the compatibility of technologies with organizational needs, scalability, and long-term viability aid in making informed technology adoption decisions (Koul and Eydgahi, 2017).

Capability Maturity Model Integration (CMMI): This is a well-established framework that provides organizations with a structured approach to assessing and improving their processes (Pane and Sarno, 2015).

- **Process Maturity Assessment**: CMMI offers a systematic method for evaluating the maturity of an organization's processes. By assessing various process areas, organizations can gain insights into their current capabilities and identify areas for enhancement.
- **Continuous Improvement Path**: CMMI provides a defined path for continuous improvement. Organizations progress through maturity levels, from initial ad hoc processes to well-defined, managed, and optimized processes. This progression enables organizations to establish a roadmap for advancing their digital capabilities over time.
- **Best Practices Integration**: CMMI integrates industry best practices, ensuring that organizations adopt proven methodologies for process improvement. This integration is particularly valuable in the context of digital transformation, where adherence to best practices can enhance the efficiency and effectiveness of the transformation efforts.
- **Benchmarking and Goal Setting**: CMMI facilitates benchmarking against industry standards and peers. Organizations can set realistic goals based on their current maturity levels and benchmarked data, providing a clear direction for improvement in digital capabilities.
- **Risk Management**: The risk management component of CMMI is crucial for digital transformation initiatives. By systematically identifying, analyzing, and mitigating risks, organizations can navigate potential pitfalls and uncertainties that may arise during the digital transformation journey.

5.9 FUTURE TRENDS FOR END-TO-END INTEGRATION/ AUTOMATION

Anticipating the future trends in transforming business processes with end-to-end automation and low-code platforms is essential for organizations seeking to stay ahead in a dynamic and ever-evolving landscape. The convergence of these technologies is likely to shape the future of business operations in several impactful ways.

Advanced Artificial Intelligence Integration: The future will witness deeper integration of advanced AI capabilities into end-to-end automation and low-code platforms. This evolution will enable more intelligent automation, including enhanced decision-making, predictive analytics, and the ability to handle complex tasks traditionally reserved for human intervention.

Democratization of Development: As low-code platforms continue to evolve, the trend of democratizing development will become more pronounced. Non-technical users will increasingly contribute to application development, fostering a culture of innovation and agility within organizations. This shift empowers a broader range of employees to actively participate in shaping and optimizing digital processes.

Hyper-Automation and Orchestration: The concept of hyper-automation, encompassing the integration of various technologies like RPA, AI, and machine learning, will gain prominence. Organizations will focus on orchestrating these technologies seamlessly to create more sophisticated and interconnected automated workflows that span across entire business processes.

Intelligent Process Mining: Intelligent process mining, driven by AI and machine learning algorithms, will become integral for organizations seeking to understand, analyze, and optimize their business processes. This data-driven approach allows for real-time insights into process performance, leading to more informed decision-making and continuous improvement.

Integration of Blockchain Technology: The adoption of blockchain technology is poised to increase in the context of business process transformation. Particularly for industries where secure and transparent transactions are critical, integrating blockchain into end-to-end automation can enhance data integrity, security, and traceability across the entire process chain.

Sustainability Integration: Future trends in business process transformation will increasingly incorporate sustainability considerations. Organizations will leverage automation and low-code platforms to optimize processes with a focus on resource efficiency, waste reduction, and environmentally conscious practices, aligning with broader sustainability goals.

Continuous Adaptation to Regulations: As regulatory landscapes evolve, organizations will need to continuously adapt their automated processes to stay compliant. Future trends will see a heightened emphasis on building flexibility into automated workflows, allowing organizations to quickly respond to regulatory changes without significant disruptions.

5.10 CONCLUDING COMMENTS

Finally, it is important to highlight that digital transformation is a strategic imperative for organizations seeking to thrive in an increasingly digitized world. The integration of low-code and end-to-end automation technologies can significantly enhance operational efficiency, reduce costs, and improve agility. A key aspect of this transformation is the shift from product-centric to people-centric, placing customers and employees at the heart of business strategies. The integration of customer-centric and employee-centric processes goes beyond mere coexistence—it signifies a strategic alignment that permeates every aspect of organizational functioning. By embracing a culture of innovation and continuous learning, businesses can harness the full potential of end-to-end automation and low-code platforms, driving sustainable growth and competitiveness in the digital era, while also enhancing the overall customer and employee experience.

REFERENCES

Alsaadi, H. A.; Radain, D. T.; Alzahrani, M. M.; Alshammari, W. F.; Alahmadi, D. and Fakieh, B. (2021). Factors that affect the utilization of low-code development platforms: Survey study. *Romanian Journal of Information Technology and Automatic Control*. 31(3), 123–140.

Alsufyani, N. and Gill, A. Q. (2021). A review of digital maturity models from adaptive enterprise architecture perspective: Digital by design. *2021 IEEE 23rd Conference on Business Informatics (CBI)*, Bolzano, Italy, pp. 121–130. doi: 10.1109/CBI52690.2021.00023

Chaudhari, A. R. and Joshi, S. D. (2021). Study of effect of agile software development methodology on software development process. *2021 Second International Conference on Electronics and Sustainable Communication Systems (ICESC)*, Coimbatore, India, pp. 1–4. doi: 10.1109/ICESC51422.2021.9532842

Doxey, C. H. (2019). *Internal Controls Toolkit*. Wiley, New Jersey.

Fischer, G.; Lundin, J. and Lindberg, J. O. (2020). Rethinking and reinventing learning, education and collaboration in the digital age – From creating technologies to transforming cultures. *International Journal of Information and Learning Technologies*. 37(5), 241–252.

Harmon, P. (2003). *Business Process Change. A Business Process Management Guide for Managers and Process Professionals*. 4th Edition. Morgan Kaufmann Publishers, USA.

Heavin, C. and Power, D. J. (2018). Challenges for digital transformation – Towards a conceptual decision support guide for managers. *Journal of Decision Systems*. 27(sup 1), 38–45.

Kaaria, E. and Shamsuzzoha, A. (2024). Improvement of an order-to-cash business process by deploying lean six sigma tools: A case study. *International Journal of Productivity and Performance Management*. 73(11), 161–189.

Koul, S. and Eydgahi, A. (2017). A systematic review of technology adoption frameworks and their applications. *Journal of Technology Management & Innovation*. 12(4), 106–113.

Kraus, S.; Durst, S.; Ferreira, J.J.; Veiga, P.; Kailer, N. and Weinmann, A. (2022). Digital transformation in business and management research: An overview of the current status quo. *International Journal of Information Management*. 63, 102466.

Lameijer, B. A.; Boer, H.; Anthony, J. and Does, R. J. M. M. (2023). Continuous improvement implementation models: A reconciliation and holistic metamodel. *Production Planning & Control*. 34(11), 1062–1081. doi: 10.1080/09537287.2021.1974114

Leal-Rodríguez, A. L.; Sanchís-Pedregosa, C.; Moreno-Moreno, A. M. and Leal-Millán, A. G. (2023). Digitalization beyond technology: Proposing an explanatory and predictive model for digital culture in organizations. *Journal of Innovation & Knowledge*. 8, 100409.

Li, S.; Gao, L.; Han, C.; Gupta, B.; Alhalabi, W. and Almakdi, S. (2023). Exploring the effect of digital transformation on firms' innovation performance. *Journal of Innovation and Knowledge*. 8(1), 100317.

Martinez, E. and Pfister, L. (2023). Benefits and limitations of using low-code development to support digitalization in the construction industry. *Automation in Construction*. 152, 104909.

Matt, C.; Hess, T.; Benlian, A. (2015). Digital transformation strategies. *Business and Information Systems Engineering*. 57(5), 339–343. http://link.springer.com/article/10.1007/s12599-015-0401-5

Pane, E. S. and Sarno, R. (2015). Capability maturity model integration (CMMI) for optimizing object-oriented analysis and design (OOAD). *Procedia Computer Science*. 72, 40–48.

Phillips, J. and Klein, J. D. (2023). Change management: From theory to practice. *TechTrends*. 67, 189–197. doi: 10.1007/s11528-022-00775-0

Reis, J.; Amorim, M.; Melão, N. and Matos, P. (2018). Digital transformation: A literature review and guidelines for future research. In Á. Rocha; H. Adeli; L. P. Reis and Costanzo, S. (eds.). *Trends and Advances in Information Systems and Technologies*. WorldCIST'18 2018. Advances in Intelligent Systems and Computing. 745, 411–421. Springer, Cham.

Rokis, K. and Kirikova, M. (2022). Challenges of low-code/no-code software development: A literature review. In È. Nazaruka; K. Sandkuhl and U. Seigerroth (eds.). *Perspectives in Business Informatics Research*. BIR 2022. Lecture Notes in Business Information Processing. 462, 3–17. Springer, Cham.

Sanchis, R.; García-Perales, Ó.; Fraile, F. and Poler, R. (2020). Low-code as enabler of digital transformation in manufacturing industry. *Applied Sciences*. 10, 12.

Serrala (2019). *Cognitive O2C Management*. Whitepaper, Serrala.

Shah, D.; Rust, R. T.; Parasuraman, A.; Staelin, R. and Day, G. S. (2006). The path to customer centricity. *Journal of Service Research*. 9(2), 113–124.

Zaoui, F. and Souissi, N. (2020). Roadmap for digital transformation: A literature review. *Procedia Computer Science*. 175, 621–628.

Zhang, S.; Fan, D.; He, J. and Zhang, P. (2021). A new approach for end to end automation testing platform with cloud computing for 5G product. *2021 International Conference on Computer Engineering and Application (ICCEA)*, Kunming, China, 2021, pp. 322–326. doi: 10.1109/ICCEA53728.2021.00070

Zhu, X.; Ge, S. and Wang, N. (2021). Digital transformation: A systematic literature review. *Computers & Industrial Engineering*. 162, 107774.

6 Improving Decision-Making with Business Intelligence

A Case Study of Lusilectra

Daniel Freitas and Adelaide Martins
University of Porto, Porto, Portugal

Vítor Correia
Lusilectra, Porto, Portugal

6.1 INTRODUCTION

Contemporary organizations operate in a socio-economic reality characterized by intense competition, continuous change, and increasing uncertainty (Bundy et al., 2017; Kahn et al., 2013; Valeri, 2021). Changes brought about by new technologies and disruptive events such as the COVID-19 pandemic affect organizations of all sizes, profiles, and fields of activity (Burlea-Schiopoiu et al., 2023). In this context, the survival and prosperity of organizations depend on their ability to respond quickly and assertively to all these changes and to anticipate market shifts.

The development of a digital transformation strategy should focus on reducing the risk of an organization becoming obsolete and maximizing the benefits derived from the use of new digital resources (Burlea-Schiopoiu et al., 2023). Therefore, it is essential that decision-makers have access to high-quality, up-to-date, organized, and timely information to make decisions that ensure the success of the organization. In this context, business intelligence (BI) solutions become particularly valuable as they enable the reporting, processing, and visualization of the best possible information. The implementation of BI solutions requires information that depends on the prior implementation of robust enterprise systems (ES).

Burlea-Schiopoiu et al. (2023) state that the improvement of digital competencies is an ongoing process, unaffected by disruptive elements such as crises but shaped by emerging technologies. The COVID-19 crisis has accelerated the development and deployment of new technologies, and consequently, the pandemic crisis has had a direct and positive impact on how organizations develop and implement strategies for digital transformation and the adoption of new digital technologies. New digital developments are contributing to the evolution of the ES that are enabling even more comprehensive and real-time data collection, enriching the analytical capabilities of BI, and improving strategic decision-making.

 DOI: 10.1201/9781003461821-6

Over the past few decades, organizations, including small and medium enterprises, have accumulated vast amounts of data that, if not properly processed, is useless or of limited use to decision-makers (Kimball & Ross, 2013). Data that cannot be transformed into information and knowledge is a major challenge for organizations and prevents them from gaining competitive advantage (Brands & Holtzblatt, 2015).

BI solutions enable organizations to extract value from the vast amount of existing data. Abai et al. (2019) define BI as a collection of technologies, applications, and practices typically used to support business decisions. Mishra and Saini (2018, p. 405) state that "BI is an approach that makes use of technology to assist the organisation and achieve their long-term goals by aligning the business with the organisation strategy". Indeed, the use of BI provides decision-makers with timely access to information, allowing them to extract critical facts from corporate data in real-time or near real-time. As a result, decision-makers become more effective in their decisions, which improves the performance of the organization (Kimball & Ross, 2013).

Management accounting serves as the primary information framework that supports decision-making and control processes in many organizations (Rikhardsson & Yigitbasioglu, 2017). It represents the process of identifying, measuring, accumulating, analysing, preparing, interpreting, and communicating information (Rikhardsson & Yigitbasioglu, 2017). Information is used by management to plan, evaluate, and control an organization and to ensure the correct and responsible use of its resources (Atkinson et al., 2011; Rikhardsson & Yigitbasioglu, 2017). The use of key performance indicators (KPIs) is a key factor in the effective application of management control mechanisms. This expansion of the role of management control in organizations is inextricably linked to technological developments and the increasing digitalization of processes (Rikhardsson & Yigitbasioglu, 2017).

This chapter explores the importance of BI for decision-making, through the preparation of a BI solution in Lusilectra, a company that is part of the Salvador Caetano Group (GSC), a leading group in the automotive retail sector in the Iberian Peninsula. From this objective, the following specific objectives were defined: 1) analyse the motivations for using this type of solution; 2) select the KPIs to be used; and 3) identify the main challenges expected in the implementation of the BI solution.

Lusilectra has existed for over 40 years and has operated in several different business areas, so it is crucial to pay attention to its internal processes and competition. It also offers after-sales services. This multiplicity of business areas and the provision of after-sales services, in a highly competitive market, require additional planning and control efforts.

The chapter proceeds as follows: Section 6.2 discusses the most relevant literature and concepts. Section 6.3 presents the methodological approach. Section 6.4 provides a brief overview of Lusilectra. Section 6.5 offers a discussion of the main findings, and Section 6.6 contains the concluding remarks.

6.2 LITERATURE REVIEW

6.2.1 The Importance of Management Control in Decision-Making and the Key Performance Indicators

In the past, there has been an excessive focus on financial controls and a separation between management control and the strategic and operational components

(Langfield-Smith, 1997). The idea of management control as a mechanism for monitoring budget execution has become outdated, and it has been identified as a bridge between strategic planning and operational control (Fähndrich, 2023). Anthony and Govindarajan (1995) support this view making a distinction between strategic planning, which involves the definition of objectives, strategies, and policies, and operational control, which ensures the effectiveness and efficiency of tasks. Management control is the link between strategic planning and operational control and is realized by translating long-term strategic goals into short-term operational objectives and actions for the organization (Anthony & Govindarajan, 1995). Thus, control mechanisms are no longer merely diagnostics but have a more proactive rather than reactive role, directing behaviour in the direction desired by the organization and ensuring that it is consistent with the defined objectives and strategies (Kaplan & Norton, 1996).

Management control plays an important role in organizational decision-making and is closely related to technological improvements (Rikhardsson & Yigitbasioglu, 2017). In fact, digital transformation promotes more reliable and centralized data, making it easier to obtain information. It is undeniable that the use of BI solutions has become essential to provide decision-makers with information that enables better decision-making (Appelbaum et al., 2017) while also contributing to the definition of KPIs.

KPIs represent a set of measures focused on the performance of organizations that are critical to their success in the short, medium, and long term (Domínguez et al., 2019; Parmenter, 2015). They contribute to the translation of the strategic objectives of organizations into reality, making it possible to compare the results with the objectives set by the organizations and to identify deviations (Domínguez et al., 2019; Popova & Sharpanskykh, 2010). As such, they make it possible to measure the internal processes of organizations and contribute to their level of performance (Kaplan & Norton, 1996).

Defining KPIs is a complicated task because it includes many aspects to consider, such as business strategy, business objectives, measurement, analysis, and reporting (Domínguez et al., 2019). According to Eckerson (2009), the creation of KPIs is a mixture of art and science. The author suggests a set of rules for creating KPIs. The first of these is "less is more" because many KPIs do not have a major impact on organizational performance. Usually, only a very small number of KPIs have a high impact on results. Most people cannot focus on multiple KPIs at the same time. Focusing on a few KPIs gives stakeholders a deeper understanding of the behaviours they need to follow. Using a broad set of KPIs requires additional effort and costs that may not add value.

KPIs should be extensible by function, level, and task. It is advisable to use a reduced set of KPIs that can be broken down into more detailed KPIs according to the needs of their users. For example, executives should have a compact dashboard with strategic KPIs but with the ability to drill down and observe the operational KPIs that influence these indicators. KPIs should be simple and easy to understand, and it is crucial to understand both what is being measured and how the calculations are made. Overly complex KPIs are difficult to understand and will not drive behaviour in the desired way, thus damaging performance. Therefore, KPIs should be easy to understand and action-oriented. Users should know how to act on the results presented by the KPIs. To achieve this, they need to understand the link between the KPIs and the strategic objectives that created them, so that they know the correct course of action in the event of significant deviations.

All KPIs should have an owner to ensure accountability by being assigned to a person or team (Eckerson, 2009). This accountability leads to greater motivation for more diligent execution of the KPIs. Data credibility is also important, and experience shows that if users do not trust the data, they will not use it. The data must be accurate, with information about its origin, and there must be no doubt about its veracity.

Another important feature is the correlation between different KPIs (Eckerson, 2009). Since they are defined according to the strategy, there must be a cause–effect relationship between them. Knowing that each strategic objective is made up of one or more KPIs, it is possible to identify the impact that the performance of one or more KPIs has on the others. For example, this relationship allows organizations to identify the KPIs that are driving the strategy if they are not meeting their objectives. KPIs should measure the performance of organizations in multiple business dimensions, not just financial. They must be aligned with the strategy and should be consistent among them. For KPIs to be effective, users must be involved in their definition, which increases user engagement and knowledge and reduces the likelihood of "manipulating" the system.

Decision-making is extremely important for achieving strategic and operational objectives in any organization (Estrada-Torres et al., 2018). Hence, decision-making must be guided by relevant information. Decision-making based on KPIs has been recognized as an important aspect in organizations, as it helps to measure the progress made towards their objectives (Nura & Osman, 2012). The quality and speed of decisions can also be a success factor for organizations.

The KPIs that enable the performance monitoring depend on data, which can be financial or non-financial. Initially, organizations were prepared to produce only financial information, which resulted in limited mechanisms for gathering and analysing non-financial information (Ittner & Larcker, 2003). Without this information, organizations faced significant challenges in defining and monitoring KPIs that required more detailed information. As a result, organizations had to adapt their internal processes to collect the data needed to obtain non-financial information. Technological limitations have been the main barriers to obtaining this non-financial information, but the evolution of information systems has enabled an exponential increase in the data available. BI has made it possible to collect and transform data into useful information for defining and monitoring KPIs and improving the effectiveness of decision-making (Rikhardsson & Yigitbasioglu, 2017).

6.2.2 The Benefits of Business Intelligence and the Critical Success Factors for Its Implementation

The aim of BI systems is to make information easily accessible, understandable, and intuitive for decision-makers (Kimball & Ross, 2013) through the process of exploring, integrating, and analysing information from different data sources in a multidimensional way (Olszak & Ziemba, 2007).

The use of a BI solution provides higher quality information that assists decision-makers in the decision-making process, monitoring KPIs, and following the organization's strategy (Watson & Wixom, 2007). According to Williams and Williams (2007), the use of a BI solution makes it possible to perform analyses related to the profitability of the marketed products and services (e.g., information on sales volume, sales costs, and margins obtained), customer relations (e.g., information on increasing market share, level of satisfaction; customer retention and customer location and profitability), management control (e.g., information on deviations from budget, income and expenditure trends, cash requirements, and identification of customer balances with relevant seniority), and strategic planning (e.g., monitoring the organization's KPIs and its strategy). Watson and Wixom (2007) state that a BI solution saves controllers time because the more efficient delivery of data eliminates the need for manual data collection.

BI is closely related to data usage. Thus, the intertwined concepts of data governance and data culture emerge as crucial issues for any BI solution implementation. Data governance is a company-wide formal framework for effectively managing data as a corporate asset by assigning specific rights and responsibilities (Otto, 2011). The primary motivation for data governance is to treat data as a valuable corporate asset (Panian, 2010), which is consistent with data culture that refers to the collective behaviours, beliefs, and practices regarding the value and importance of data across the organization. As the data culture grows and the maturity of BI solutions increases, users are starting to look for predictive analytics that can be provided by business analytics solutions. With this evolution, organizations are beginning to make strategic decisions supported by these solutions, such as entering new markets and launching new product lines.

Critical success factors (CSFs) in the implementation of a BI system consist of a set tasks and procedures that the organization must follow to ensure the success of the project (Olszak, 2016; Scholz et al., 2010). BI solutions must be flexible and adaptable to changes in the organization's business to keep up with the new information needs of decision-makers (Yeoh & Koronios, 2010). The development cycle for BI solutions is iterative and unfinished. Therefore, BI solutions need to be continuously developed and optimized, and their success is achieved when users consistently produce knowledge from the information provided (Olszak & Ziemba, 2007).

Mungree et al. (2013) identified a set of CSFs for implementing a BI solution: 1) **Executive support and commitment**. Several authors argue that BI solution implementation initiatives should be *top-down* (e.g., Díaz et al., 2018; Olszak & Ziemba, 2007; Watson & Wixom, 2007; Yeoh & Koronios, 2010), and it is essential to have the human and financial resources needed for a successful implementation. 2) **Skills of the elements involved**. They must have the knowledge necessary for implementation. The users of these solutions must also have the necessary skills, and organizations must assess and promote training to enable the acquisition of these skills. 3) **Flexible and appropriate technological framework**. Technological solutions must ensure the quality and integrity of data while adapting to changes in the business (Yeoh & Koronios, 2010; Olszak & Ziemba, 2007). This flexibility enables the relationship between changing information needs and the *hardware* and *software* used in BI solutions. 4) **Aligning BI solutions with strategy**. It is essential to align

BI solutions with the strategy of organizations and their business orientation (Yeoh & Koronios, 2010). Otherwise, they will not provide the right KPIs to their users, losing their interest, usefulness, and relevance. 5) **Well-defined BI solution requirements**. Without this, it is not possible to ensure the consistency and integration of information (Yeoh & Koronios, 2010) and the ability to make decisions based on BI solutions is compromised. 6) **User-centric change management**. It is important to have a policy of continuous training in data literacy and to promote a culture of data democratization and data-driven decision-making (Yeoh & Koronios, 2010). To this end, future users must be involved in the various stages of the implementation process so that the solutions are aligned with their needs (Olszak & Ziemba, 2007). 7) **Effective data management**. Data reliability must be ensured, and data must be up-to-date, complete, and standardized so that BI solutions can support effective decision-making (Watson & Wixom, 2007). 8) **Project owner**. There must be one person responsible for the implementation of the project, someone who understands the business and technology components and can facilitate the connection between the two (Yeoh & Koronios, 2010). 9) **Project scope management**. Organizations must not let the scope of the project wane and must try to test the developments made (Olszak & Ziemba, 2007). This procedure helps to see the benefits of the developed BI solutions and allows for constructive *feedback* between the team developing the solution and its future users (Yeoh & Koronios, 2010).

6.3 METHODOLOGY

This research follows a qualitative research approach based on a single case study – Lusilectra, a company of a leading group in the automotive retail sector. This study aims to explore the importance of BI for decision-making, through the preparation of a BI solution in Lusilectra. From this general objective, the aim is to delve deeper into the analysis of the motivations for using this type of solution, the selection of the KPIs to be used, and the identification of the main challenges that the company perceives that it will face in the implementation process.

Data sources include semi-structured interviews, document analysis, and participant observation. The preparation of the semi-structured interviews is based on the literature review and the research objectives (Mahama & Khalifa, 2017). Interviews were conducted with the Executive and Financial Administrator (EFA), the Executive and Operational Administrator (EOA), two directors (D1 and D2), each one of them responsible for different business areas, and the Information Technology (IT) Professional (ITP), an employee of the IT department of the GSC. The interviewees were purposively selected based on their involvement and relevance in the BI implementation process. The interviews were conducted online or in person during the second quarter of 2023 and were recorded with the prior permission of the interviewees and transcribed verbatim for qualitative analysis.

According to the guidelines for case-based research, other documents and supplementary data were collected to understand the context in which the preparation of the BI solution takes place at Lusilectra and the path that has been developed so far. The data includes Lusilectra's annual reports, Lusilectra's annual budget reports, strategic management reports, Lusilectra's website, internal email and Microsoft

Teams communications, management maps, BI solutions developed by GSC, data available in the enterprise resource planning (ERP), maps and reports from the quality team, presentations to management, minutes of general meeting, and X matrices and Kaizen boards.

Meetings were also held with the business partner responsible for the ERP used and with other companies in the group that had already implemented BI solutions, as well as with the group's planning, management control, and internal audit departments. One of the authors carried out this research as an insider of Lusilectra. He worked as a controller in the company and is the owner of the implementation process of the BI solution.

6.4 A BRIEF OVERVIEW OF LUSILECTRA DIGITAL AND MANAGEMENT TOOLS

Lusilectra is dedicated to the sale of products and services for automotive workshops, cargo handling, and accessories for light and heavy vehicles (Lusilectra, 2023). The company has offices in Porto, Lisbon, and Madrid and operates in France, Angola, Cape Verde, Mozambique, and Senegal. It has four business areas with different characteristics, challenges, and specificities: accessories, forklifts, equipment, and tools. These areas provide an after-sales service that guarantees the installation, maintenance, and repair of the products sold.

In terms of its digital transformation, the company already has had an ERP, the PHC CS Enterprise, for more than a decade. Until then, the company benefited from the IT solutions available to the entire GSC. However, these solutions did not efficiently and predictably meet the needs of the business areas (PHC Software, 2023). Beyond the unique features and functionalities of each ES, the ERP system facilitates the collection, processing, and storage of data that becomes information available for later use by the organization in decision-making. In fact, the ERP system was customized according to the characteristics of the existing business areas. Such an ES is key to the successful implementation of a BI solution due to the huge amount of historical data it contains. It has also been instrumental in centralizing and integrating information that is useful to the company.

Lusilectra also uses several strategic management tools to define and monitor its objectives, allowing the company's administrators and directors (henceforth top management) to align strategic and operational KPIs. The company uses the Hoshin Kanri X Matrix. This methodology helps organizations with strategic planning, change management, and ensuring that the direction, goals, and objectives are correctly defined, communicated, monitored, and corrected through the feedback received (Souad et al., 2017). There is also a performance evaluation tool called Performance and Development Management (GDD). This tool, developed by the GSC, was implemented to measure the performance of individuals and teams. It supports the implementation of the strategic objectives. The GDD establishes individual and team KPIs to assess individual and collective performance. In addition to the KPIs derived from the X Matrix, various teams have added KPIs to measure their performance. Each year, variable remuneration is awarded to each employee based on individual and collective performance and the targets set. GSC and Lusilectra are also influenced by the Kaizen philosophy, with the aim of continuously improving the products sold, the services provided, and the internal processes. Several KPIs have been defined to monitor these improvements.

The existence of numerous KPIs resulting from different management tools makes the monitoring process difficult. Non-financial KPIs are monitored quarterly and/or annually, as there are not sufficient resources for daily and/or monthly monitoring. As a result, existing management control is primarily focused on monitoring financial indicators and does not yet provide the necessary information for fast and effective decision-making. A new administration has increased the need for a solution capable of providing updated information on the company's performance. To achieve an effective implementation of a BI solution, there is an opportunity to leverage the know-how of other group companies that already use this type of solution and to use the services of a group company specialized in IT that can provide technical support for development and implementation.

6.5 FINDINGS AND DISCUSSION

6.5.1 Motivations for Adopting a BI Solution

It is important to assess whether Lusilectra's current motivations for adopting a BI solution are aligned with the benefits outlined in the literature. The different characteristics of Lusilectra business areas and the different types of after-sales services make it very difficult to monitor performance. These factors increase the complexity of the processes and make decision-making difficult. The implementation of a BI solution at Lusilectra is supported by top management. The EFA and EOA stated that the implementation of BI was a priority. The directors interviewed confirmed this position:

> It's always beneficial to have a BI solution [...]. Because if we have one, we can improve our planning, we can control our risk better, we can have optimized routines in data collection, we can make decisions quickly and assertively, we can even improve the relationship with our customers when thinks are not going well [...], we can reduce costs and I think we can have continuous and sustainable growth.
>
> (D1)

> You [...] count the time I spend at the moment doing Excel sheets and correcting all the data to see that this tool is completely justified. [...] Now, I have time to do other things that are also important.
>
> (D2)

The administrators identified several reasons for implementing the BI solution. EFA emphasized the importance of receiving information in real-time and the need to have a clear view of the company's situation. This opinion was shared by the EOA, who considered it essential to make information available quickly.

This value that the administrators expected from the implementation of a BI solution is in line with the benefits identified in the literature (Olszak & Ziemba, 2007; Watson & Wixom, 2007).

Both administrators stated that the information must be relevant and easy to interpret. EFA mentioned that priority should be given to using dashboards and forms of visualization that make the information easy to understand, even for other people who are not familiar with the technical or financial terms. He added that the BI solution could also help employees to better understand the impact of their work on strategic objectives of the company. Improving the flow of information and increasing transparency are mentioned by Mungree et al. (2013) and Olszak and Ziemba (2007) as potential benefits of using a BI solution. According to the EOA, once these characteristics were met, the conditions were in place for faster and more sustainable decision-making. The aim was for the company to be able to keep up with the speed of business and be able to make the best decisions. According to Mungree et al. (2013), Olszak and Ziemba (2007), and Watson and Wixom (2007), the ability to make better decisions through the insights provided by the BI solution is one of the benefits of using it.

The managers and directors of Lusilectra agreed that improving the decision-making process would boost the performance of the various business areas and make the company more competitive. For D1, this performance improvement was not limited to profit analysis. Obtaining information on the inventory turnover of the business area he managed and the productivity of the after-sales service technicians was important for optimizing resources. This view was supported by the literature (e.g., Williams & Williams, 2007), which suggests that a BI solution can deliver tangible, measurable benefits such as reducing labour costs and improving operational processes. Olszak and Ziemba (2007) and Watson and Wixom (2007) add that analysing business costs is an important benefit of using a BI solution.

One of the aspects most highlighted by the interviewees is the need to obtain information about the operational process of the business areas. There were difficulties in obtaining information on the control of operations, mainly due to the need for managers to carry out their analysis to monitor the factors they consider relevant. Mungree et al. (2013) and Watson and Wixom (2007) stated that a BI solution saves time in providing information to stakeholders, allowing for quick and effective analysis. For Olszak and Ziemba (2007), this type of solution overcomes the limitations of traditional information management systems and provides added value to decision-makers. EFA underlined the importance of this point with the difficulty in finding the scattered data stored in the ERP, spreadsheets, and other databases. The data needed to compile information was scattered, and there was no single source of truth across the company. The use of a BI solution makes it possible to gather, process, and analyse large amounts of data from different sources (Olszak & Ziemba, 2007) and overcomes the limitations and difficulties caused by data dispersion. EFA added that the BI solution should reveal the interconnection between different KPIs, allowing a holistic view of the company's performance and identifying areas for improvement. This helps to understand how each KPI fits into the overall strategy. A BI solution makes it possible to better explore the different variables of business performance, providing more comprehensive information from different perspectives (Olszak & Ziemba, 2007; Watson & Wixom, 2007).

D2 suggested that the use of BI speeds up data collection and processing, reducing the likelihood of errors and increasing the accuracy of analysis. This finding is consistent with the argument put forward by Olszak and Ziemba (2007), who advocate reducing the time to analyse information and automation as motivations for BI implementation. D2 also believed that the BI solution would allow them to identify areas of risk and take action in advance. Minimizing risk is cited by Mungree et al. (2013) as one of the benefits of a BI solution, as it allows companies to respond quickly to changes in financial conditions or customer preferences.

Overall, the evidence suggests that the main motivation for implementing a BI solution at Lusilectra was to improve decision-making at strategic and operational levels. It was essential to speed up access to relevant information to monitor the activities of the business areas and improve their performance. To this end, it was important to improve the process of collecting, processing, and presenting data, making it easier to read. Other aspects such as speed of access to information, increased transparency of information, involvement of employees in the decision-making process, and the ability to relate different indicators to the company's strategy were highlighted.

However, some of the motivations identified in the literature were not observed in the interviews. Information security and predictive analytics, mentioned by Watson and Wixom (2007), were not mentioned by the interviewees. Monitoring the competition, combining internal company data with data from the external environment (statistical portals, financial portals, and other sources), and identifying commercial fraud, highlighted by Olszak and Ziemba (2007), were also not considered to be issues of high priority when implementing a BI solution.

6.5.2 Selection of Key Performance Indicators

The selection of the KPIs is a key step in the successful implementation of the BI solution (Yeoh & Koronios, 2010). Based on Lusilectra's existing strategic management tools, a set of KPIs was selected for the first version of the BI solution. In this respect, the EOA supported the decision autonomy of the controller in the selection of the KPIs: "It is he [controller] who, in essence, must try to translate into a BI solution what are the concerns and [...] the vision of the management in terms of indicators".

A reduced set of strategic KPIs with the greatest impact on decision-making was selected, avoiding the dispersion of the analysis with more operational KPIs. Simplicity and ease of interpretation were factors influencing this selection. The KPIs should allow analysis by business area and distinguish between commercial and after-sales. Strategic alignment was ensured in this selection, as all the KPIs chosen for the BI solution were previously defined by Lusilectra top management and their accountability was guaranteed. Data credibility was also a priority, ensuring that the data feeding these KPIs was reliable and transparent. Any KPI that was not measured by properly validated and automated data, or whose data came from unreliable sources, was excluded from the solution. The correlation between the KPIs and their coherence with the company's strategy were factors considered in the selection, in order to avoid isolated analyses. Finally, the origin of the data also influenced the choice: the KPIs obtained through the ERP were prioritized. Eleven KPIs could be obtained by integrating ERP data into the data warehouse. Only one of them – the number of Kaizen ideas implemented – does not come from this source and is intended to test the ability of the IT teams to obtain data from other sources, transform it, and incorporate it into the data warehouse. Accordingly, the KPIs defined for the implementation of the BI solution are the following:

a. **Turnover** allows to understand the sales performance of the company's business areas, their growth, and their market positioning to analyse their productivity. In order to identify opportunities and actions for improvement, it is important to disaggregate this KPI: 1) by customer; 2) by country, to identify turnover by market and making it possible to adjust expansion strategies; 3) by region, in the markets with the greatest presence; 4) by salesperson, to analyse their individual performance and productivity and making it possible to adjust incentives; and 5) by product, to identify the best performing products and those that require further attention or intervention (e.g., marketing campaigns or discounts).
b. **Gross margin** provides an understanding of the profitability of the business units and allows the implementation of profit improvement strategies to be assessed. This KPI is presented to top management monthly. All the information is available in the ERP system. In order to identify opportunities for optimization, this KPI needs to be broken down: 1) by customer, to identify the most and least profitable and adjust the commercial strategy; 2) by country, to identify variations in gross margins due to economic, cultural, or regulatory factors, in order to exploit opportunities or reduce exposure to markets with a downward economic cycle; 3) by region, to identify the profit of the regions in the markets with the greatest presence; 4) by salesperson, making it possible to assess the profitability of the salesperson; and 5) by product, to identify the contribution of the products sold to the increase in the gross margin and to identify those that require a review of the pricing strategy.

   Gross margin should be analysed in conjunction with turnover and other indicators. The BI solution should present a comparative view. Indeed, a business area strategy may prioritize the development of only one of these KPIs (e.g., in times of expansion, sales growth is prioritized instead of gross margin). It is important to understand how these KPIs affect each other. This real-time information allows the commercial strategy to be adapted more quickly.
c. **The profitability of maintenance contracts**: Maintenance contracts are related to customer loyalty and satisfaction with after-sales service. Monitoring profitability is crucial because it is necessary to have both competitive and profitable values. This KPI is very important for after-sales service results. There was enough data in the ERP to calculate its profitability. In the BI solution, this KPI cannot be restricted to profitability. It is important to show more details: the date, the amount, and the costs already charged. This makes it possible to monitor the profitability of contracts and make wise decisions about future contracts.

d. **Technician occupancy rate** measures the allocation of after-sales service technicians during their working hours to understand if their work is being properly monetized. Service performance depends on the technicians. Their under-utilization or over-utilization can cause problems. Calculating the technician occupancy rate is a time-consuming process that involves extracting data from the ERP and using a spreadsheet. Technician occupancy times are automatically recorded in the ERP via an application.
e. **Customer balance**: The BI solution should provide a view of sales and after-sales service in each business area and whether it is within or outside the agreed payment period. Customers' balances could be organized by age, in months, allowing the company to prioritize its actions of recovery money. This KPI was provided on a monthly basis. However, the ERP has enough data to show the total balance and more detailed information in the BI solution. Monitoring this KPI in real time will have a positive impact on the company's financial performance, reducing the risk of bad debts, reviewing credit policies, and negotiating payment terms. The analysis of this indicator should be linked to other KPIs, namely turnover and gross margin, as business areas may have good economic performance that does not represent good financial performance.
f. **Supplier balance**: Similar to the previous KPI, the BI solution could provide a view of sales and after-sales service. It would also be possible to understand which values are within or outside the agreed payment period. As it represents cash to be received, supplier balances could be categorized by date to facilitate cash planning. There is a total balance value that is reported, but there is more information available in the ERP that could be used in the BI solution. In fact, monitoring this KPI in near real time can have a positive impact on Lusilectra's financial performance, reducing the risk of default and contributing to having good relations with suppliers and to getting negotiating advantages.

   As a representative of important international brands, Lusilectra strived to maintain good relations with its suppliers and avoid late payments. The several business areas had different characteristics in terms of orders placed with suppliers, due to the different quantities, values, and frequency of these orders.
g. **Stock rotation** measures the efficiency with which Lusilectra manages its inventory in each business area. This KPI was not available at the time of this study. Only the total value of stock by business area was shown to managers. With the BI solution, this KPI should be constantly monitored, showing stock rotation by business area and a list of products with no outgoing movements by month. In this way, it is possible to measure the efficiency of stock management in each business area and identify the products with no demand that are negatively affecting this KPI. The ERP already has data on stock levels, cost of goods sold, and materials used, which are recorded when sales are made. With this KPI in the BI solution, it is easier to manage stock efficiently, providing important information about products that do not have the desired rotation. This efficiency can help with operations management, reducing storage costs and managing working capital requirements.

h. **Number of leads**: All contacts received through digital channels requesting information and/or quotes for products and services are considered leads. The company has invested in digital channels. This KPI makes it possible to assess in real time whether the investment made in these channels is generating new business, new customers, and sales growth. Leads are registered in the ERP, regardless of their origin. The marketing team produces monthly reports for top management. This information only provides an overview of the development of leads, leaving out all other analysis. The BI solution can provide an answer to this KPI, since the data has already been entered into the ERP. At the time of the study, it was not possible to automatically track which leads resulted in business transactions. This issue needs to be addressed in the ERP system to ensure its availability in the BI solution.
i. **Number of complaints** aims to measure the number of complaints received by Lusilectra within each period, typically on a monthly basis. These complaints may come from customers, suppliers, or other stakeholders and may be related to problems, dissatisfaction, or failures in the products and services commercialized, as well as internal matters related to failures in the company's internal processes. The number of complaints per business area could be presented, distinguishing between sales and after-sales and between internal and external complaints. Complaints received through various internal and external communication channels were recorded in the ERP. However, the lack of an intuitive or easy visualization of the number of complaints and the distinction between business areas prevents proper follow-up by management. As the data is already in the ERP system, it should be straightforward to integrate it into the BI solution. This KPI is particularly important in after-sales service, which is more prone to complaints about delays and faults.
j. **The results of satisfaction surveys**, together with the number of complaints, are essential for measuring the level of customer satisfaction. Satisfaction surveys allow for the collection of customer feedback, which, unlike complaints, is focused exclusively on customers. This KPI within the BI solution is crucial for top management to monitor customer satisfaction more regularly in each business area, distinguishing between sales and after-sales service. The results of the satisfaction surveys were not entered into a database. Instead, the quality department recorded the responses in spreadsheets, which produced a six-monthly report of the main conclusions, causing delays in the implementation of immediate and assertive measures. The company was in the process of implementing a procedure in the ERP to automatically register complaints, with the aim of assessing customer satisfaction in real time. This initiative would enable the company to determine whether it was meeting customer expectations or if improvement actions were needed. These measures should help Lusilectra to improve its image and reputation.
k. **Number of Kaizen ideas implemented** allows top management to see whether employees are committed to continuously seeking improvements in each business area. This gives top management a more detailed view, identifies the teams that are not proposing new ideas, and enables measures

to be taken to stimulate them. The implemented Kaizen ideas were registered by the employees on a platform created by the GSC. Unlike other KPIs, the number of implemented Kaizen ideas is not sourced from the ERP. This philosophy has been in place at GSC for several years. As a result, Lusilectra has been requested every six months to provide the number of Kaizen ideas implemented, which is still done manually by extracting them into a spreadsheet.

1. **Number of technical assistance requests/repair orders (TAR/RO)** represents the number of requests received by the company's after-sales services. According to EFA, it is crucial to understand which TAR/RO are ongoing within the after-sales service, categorized by business area, their duration, and the labour and material costs incurred. This data enables managers to assess the causes of prolonged TAR/RO durations and make informed decisions about their resolution. The costs were obtained by extracting data from the ERP and performing additional calculations, resulting in a time-consuming and difficult decision-making process. This KPI allows top management to access vital information on the number of TAR/RO in progress, their duration, and the value charged, as well as TAR/RO completed, their duration, and the profit generated. Together with the technician occupancy rate, this KPI provides an overview of turnover.

6.5.3 Main Challenges Foreseen in the BI Solution Implementation

The main challenges anticipated in the implementation of the BI solution at Lusilectra are analysed according to the CSFs listed in the literature. Implementing a BI solution involves critical challenges if they jeopardize the success of the implementation (Mungree et al., 2013; Yeoh & Koronios, 2010). For ITP, the main challenge was to choose the right technological tools to respond to the company's specific problems. Integrating data from the different sources available was also considered a challenge. For this reason, ITP considers it important to use the appropriate tools and techniques for extracting, transforming, and loading (ETL) the data into the data warehouse. This is essential to ensure the quality and consistency of the data, establish measures to verify its integrity, and implement alerts to detect inconsistent data. These measures and alerts played a fundamental role in ensuring the reliability of the data, as ITP believed that the success of the implementation depended on the trust of users in the data used. This perspective is in line with the literature (e.g., Mungree et al., 2013; Olszak & Ziemba, 2007; Watson & Wixom, 2007). The integration of the BI solution with Lusilectra's existing ES and processes was very important. For ITP it was essential to guarantee the efficient and accurate transfer of data between the different systems, as well as the correct synchronization of information. This aspect was emphasized by the administrators as a major challenge, as the data was dispersed across various sources and systems. Efficient data management, as mentioned by Mungree et al. (2013), emphasizes that the data model should ensure data integrity, while the ETL process must guarantee data consistency and accuracy.

The support and involvement of top management were crucial for the successful implementation of the BI solution at Lusilectra. ITP also believed that it was necessary to have a sponsor with the influence and interest to drive the implementation project forward was necessary. This finding is supported by Mungree et al. (2013), who emphasize that management involvement and support are crucial to minimize user resistance to change. The interviews showed that both administrators considered the implementation of the BI solution to be a priority. The existence of a sponsor who is responsible for guiding the project and allocating the necessary resources to the project is key to successful implementation (Mungree et al., 2013).

The BI solution needs to be tailored to specific needs of the company. It was therefore important to understand the information requirements of the several business areas and to develop relevant, intuitive, and easily visualized analyses that met their specific needs. Hence, correctly identifying business problems is an aspect to consider when implementing a BI solution. Mungree et al. (2013) point out that the alignment between the BI solution and the business objectives is a CSF in its implementation. ITP added that there are problems that may not require a BI approach. In Lusilectra, there were situations where the problem might be in the processes and not in the analysis of the data. The administrators were aware of this and believed that changes might need to be made to processes and systems.

Establishing the correct starting point can also be critical to the successful implementation of the BI solution. ITP believed that the decision on where to start the implementation process should be made by top management. However, he recommended starting in an area that could promote quick and tangible results that would allow Lusilectra's administrators to see the benefits of the BI solution. During the interviews, they mentioned the importance of obtaining information quickly and easily to be motivated for the project. Interaction with users throughout the implementation process is also important to ensure that the BI solution meets their needs and expectations. The literature reinforces that the implementation of the BI solution should include user participation and feedback to facilitate its adoption and successful implementation (e.g., Mungree et al., 2013).

ITP also mentioned the importance of information security and privacy, highlighting the need for security measures to protect the company's confidential data. This may include classifying information according to confidentiality levels, restricting access to certain dashboards or sensitive data, and defining security levels to safeguard information.

Overall, the evidence is supported by the literature. Factors such as management support and commitment, the presence of a sponsor, and the alignment of the BI solution with specific business objectives were consistent with the literature (e.g., Mungree et al., 2013; Yeoh & Koronios, 2010). However, the importance of having a qualified and experienced team to implement the BI solution was not emphasized. This factor can pose challenges and potentially jeopardize the success of the implementation (Mungree et al., 2013). In addition, the ITP did not mention the importance of the technological infrastructure, particularly the hardware, focusing only on the software. However, the standardization of existing hardware within the GSC allows this professional to have in-depth knowledge of Lusilectra's hardware. This suggests that ITP may not have considered this issue to be a significant obstacle to BI implementation. The literature underscores the importance of having a flexible technology infrastructure that can meet the requirements of a BI solution

(Mungree et al., 2013). The ITP did not address the importance of a strong data-driven culture and data governance within the organization, which is not consistent with practices widely advocated in the literature.

6.6 CONCLUDING REMARKS

The main objective of implementing a BI solution at Lusilectra is to provide relevant information quickly and intuitively to facilitate a more accurate and precise decision-making process. This chapter aims to facilitate the understanding of the motivations for the implementation of a BI solution, the selection of the KPIs to be used, and the identification of the main challenges expected in the future implementation process.

The implementation of a BI solution is supported by top management and is driven by several motivations, such as the rapid availability of information, access to intuitive and easily interpretable information, improved information flow and transparency, faster and more sustainable decision-making processes, improved performance and competitiveness, resource optimization, saving time and resources in data preparation, achieving of a single source of truth, real-time strategic execution control, reducing information errors, and identifying risk areas. There is a relevant alignment between the motivations identified in this study and those found in the literature. This underlines that top management is aware of the benefits of using BI solutions and that these benefits are aligned with the needs of the organization, justifying their implementation in a company such as Lusilectra. However, some potential motivations highlighted in the literature, such as enhancing data security, providing predictive analysis, monitoring the market, comparing internal data with external sources, and identifying fraud were not explicitly mentioned.

Without the BI solution, Lusilectra would face challenges in tracking the non-financial indicators embedded in various strategic management tools. Therefore, the selection of KPIs to be included in the BI solution becomes particularly important. A limited number of KPIs have been proposed for the BI solution and have been evaluated according to Eckerson (2009). These KPIs are derived from the strategic management tools already in use at Lusilectra and ensure the existence and relevance of the information to be used in the BI solution.

In order to prepare for the future implementation of the BI solution, an effort was made to anticipate the main potential challenges that could jeopardize the successful implementation of the solution. The main challenges include selecting the appropriate tools, integrating data from several different sources within the company, ensuring the reliability and consistency of data through the ETL process, integrating the BI solution with Lusilectra's existing systems and processes, gaining the support and commitment of top management, securing a project sponsor, customizing the solution to meet the specific needs of the company, establishing internal processes to provide the necessary data for the desired analyses, defining the exact starting point, maintaining continuous interaction with users throughout the implementation process, and addressing information security and privacy concerns. Overall, there is alignment between the ITP perspective and the existing literature (e.g., Mungree et al., 2013; Olszak & Ziemba, 2007; Watson & Wixom, 2007). However, aspects such as the quality and experience of the implementation team, the technological infrastructure (hardware), and its flexibility, which are identified as CFSs in the literature, were not explicitly mentioned by ITP, but they should be considered during the preparations for the implementation of the BI solution.

This study contributes to the literature on management control, BI, and ES. The analysis of top management motivations for adopting a BI solution suggests that the evidence is supported by the existing literature (e.g., Mungree et al., 2013; Olszak & Ziemba, 2007; Watson & Wixom, 2007; Williams & Williams, 2007). The relationship between the BI solution and management control reinforces the importance of KPI selection when using a BI solution. The KPI selection process has been discussed in detail, providing a practical example of how KPIs can be selected based on pre-existing strategic management tools. However, simply selecting KPIs is not enough to implement a BI solution. The literature identifies a wide range of KPIs that present several challenges for companies to consider. This study provides valuable insights into the importance of BI solutions within the context of management practices and decision-making in the business sector of sales and technical support for automotive repair workshop equipment.

However, without the development of the implementation of a BI solution, it is impossible to understand its real impact on decision-making. In fact, a pilot implementation based on this research has been running in the company since the last quarter of 2023. A preliminary assessment with the project owner had already highlighted the critical importance of fostering a well-developed data culture within the organization, which was lacking at the time of this study and the absence of which was hampering the effectiveness of the BI efforts. As a result of this lack, it was evident that there was poorly captured and categorized data, compromising the quality of the information derived from it. This poor-quality data is the result of a lack of understanding by employees of the benefits of collecting and using data.

A comparison between the results achieved with the BI solution and what was expected is an interesting avenue for future research. In addition, analysing the alignment of the expectations that motivated top management to implement a BI solution and the selected KPIs for better decision-making with the results of the effective implementation will further enrich this field of research.

The results of this study can serve as a valuable source of information for other companies, both within and outside the business sector, providing inspiration and guidance for the development of similar solutions. The findings provide important insights into the preparation and implementation of BI solutions in other companies. The motivations, defined KPIs, and challenges can be shared with other organizations. While it is important to evaluate each specific case, these clues provide valuable contributions to the study of improving decision-making with BI solutions.

REFERENCES

Abai, N. H. Z., Yahaya, J., Deraman, A., Hamdan, A. R., Mansor, Z., & Jusoh, Y. Y. (2019). Integrating BI and analytics in managing public sector performance: An empirical study. *International Journal on Advanced Science, Engineering and Information Technology*, 9(1), 172–180. https://doi.org/10.18517/ijaseit.9.1.6694

Anthony, R. N., & Govindarajan, V. (1995). *Management Control Systems* (8th ed.). Irwin.

Appelbaum, D., Kogan, A., Vasarhelyi, M., & Yan, Z. (2017). Impact of business analytics and enterprise systems on managerial accounting. *International Journal of Accounting Information Systems*, 25, 29–44. https://doi.org/10.1016/j.accinf.2017.03.003

Atkinson, A. A., Kaplan, R. S., Matsumura, E. M., & Young, S. M. (2011). *Management Accounting: Information for Decision-making and Strategy Execution* (6th ed.). Pearson (Intl). ISBN 9780273769989.

Brands, K., & Holtzblatt, M. (2015). Business analytics: Transforming the role of management accountants. *Management Accounting Quarterly*, 16(3), 1–12.

Bundy, J., Pfarrer, M. D., Short, C. E., & Coombs, W. T. (2017). Crises and crisis management: Integration, interpretation, and research development. *Journal of Management*, 43(6), 1661–1692. https://doi.org/10.1177/0149206316680030

Burlea-Schiopoiu, A., Borcan, I., & Dragan, C.O. (2023). The impact of the COVID-19 crisis on the digital transformation of organizations. *Electronics*, 12(5), 1205. https://doi.org/10.3390/electronics12051205

Díaz, A., Rowshankish, K., & Saleh, T. (2018). Why data culture matters. *McKinsey Quarterly*, 3, 36–53.

Domínguez, E., Pérez, B., Rubio, Á. L., & Zapata, M. A. (2019). A taxonomy for key performance indicators management. *Computer Standards & Interfaces*, 64, 24–40.

Eckerson, W. W. (2009). *Performance Management Strategies: How to Create and Deploy Effective Metrics*. The Data Warehousing Institute. Retrieved from https://tdwi.org/research/2009/01/bpr-1q-performance-management-strategies.aspx

Estrada-Torres, B., Del-Río-Ortega, A., Resinas, M., & Ruiz-Cortés, A. (2018). On the relationships between decision management.

Fähndrich, J. (2023). A literature review on the impact of digitalisation on management control. *Journal of Management Control*, 34(1), 9–65. https://doi.org/10.1007/s00187-022-00349-4

Ittner, C. D., & Larcker, D. F. (2003). Coming up short on nonfinancial performance measurement. *Harvard Business Review*, 81(11), 88–95.

Kahn, W. A., Barton, M. A., & Fellows, S. (2013). Organizational crises and the disturbance of relational systems. *Academy of Management Review*, 38(3), 377–396.

Kaplan, R. S., & Norton, D. P. (1996). Using the balanced scorecard as a strategic management system. *Harvard Business Review*, *74*(1), 75–85.

Kimball, R., & Ross, M. (2013). *The Data Warehouse Toolkit* (3rd ed.). John Wiley & Sons, Inc.

Langfield-Smith, K. (1997). Management control systems and strategy: A critical review. *Accounting, Organizations and Society*, 22(2), 207–232.

Lusilectra. (2023). Apresentação. Retrieved August 24, 2023, from https://lusilectra.com/pt/empresa/apresentacao/

Mahama, H., & Khalifa, R. (2017). Field interviews: Process and analysis. In *The Routledge Companion to Qualitative Accounting Research Methods* (pp. 321–338). Routledge.

Mishra, R., & Saini, A. K. (2018). Technological capabilities impacting BI success in organisations. In *Quality, IT and Business Operations* (pp. 405–419). Springer. https://doi.org/10.1007/978-981-10-5577-5_32

Mungree, D., Rudra, A., & Morien, D. (2013). *A Framework for Understanding the Critical Success Factors of Enterprise BI Implementation*. AMCIS.

Nura, A. A., & Osman, N. H. (2012). A toolkit on effective decision making measurement in organizations. *International Journal of Humanities and Social Science*, 2(4), 296–303.

Olszak, C. M. (2016). Toward better understanding and use of BI in organizations. *Information Systems Management*, 33(2), 105–123.

Olszak, C. M., & Ziemba, E. (2007). Approach to building and implementing BI systems. *Interdisciplinary Journal of Information, Knowledge, and Management*, 2(1), 135–148.

Otto, B. (2011). Organizing data governance: Findings from the telecommunications industry and consequences for large service providers. *Communications of the Association for Information Systems*, 29, 45–66.

Panian, Z. (2010). Some practical experiences in data governance. *World Academy of Science, Engineering and Technology Management*, 62, 939–946. Parte superior do formulário" está correto? É alguma indicação interna

Parmenter, D. (2015). *Key Performance Indicators: Developing, Implementing, and Using Winning KPIs*. John Wiley & Sons.

PHC Software. (2023). Lusilectra torna-se mais flexível e ganha visão macro. Retrieved February 23, 2024, from https://phcsoftware.com/casos-de-sucesso/lusilectra/

Popova, V., & Sharpanskykh, A. (2010). Modeling organizational performance indicators. *Information Systems*, 35(4), 505–527.

Rikhardsson, P., & Yigitbasioglu, O. (2017, May). Business intelligence in management accounting research: Current status and future focus. In *40th Annual Congress European Accounting Association*.

Scholz, P., Schieder, C., Kurze, C., Gluchowski, P., & Böhringer, M. (2010). Benefits and challenges of BI adoption in small and medium-sized enterprises. *18th European Conference on Information Systems, ECIS 2010*.

Souad, D., Messaouda, C., & Maazouzi, K. (2017). Planning using Hoshin Kanri. *International Journal of Academic Research in Economics and Management Sciences*, 6(4), 1–13. https://doi.org/10.6007/ijarems/v6-i4/3175

Valeri, M. (2021). *Organizational Studies: Implications for the Strategic Management*. Springer International Publishing. https://doi.org/10.1007/978-3-030-87148-2

Watson, H. J., & Wixom, B. H. (2007). The current state of BI. *Computer*, 40(9), 96–99. https://ieeexplore.ieee.org/document/4348170

Williams, S., & Williams, N. (2007). *The Profit Impact of Business Intelligence*. Elsevier.

Yeoh, W., & Koronios, A. (2010). Critical success factors for business intelligence systems. *Journal of Computer Information Systems*, 50(3), 23–32.

7 Diagnosing the Implementation of Digital Technologies by European Union Manufacturing Multinationals

Sofia Gomes
University Portucalense and REMIT – Research on Economics, Management and Information Technologies, Porto, Portugal

João M. Lopes
Instituto Superior Miguel Torga, Coimbra, Portugal & NECE-UBI – Research Unit in Business Sciences, University of Beira Interior, Covilhã, Portugal

7.1 INTRODUCTION

The manufacturing industry in the European Union (EU) has been adopting digital technologies that include artificial intelligence (AI) (e.g. computer vision; machine learning), cloud computing (e.g. data storage; network as a service), robotics (e.g. industrial robots; cobots; robotic vision), big data analytics (BDA) (e.g. production performance analysis; predictive maintenance; supply chain optimization), high-speed infrastructures (e.g. fiber optic networks; 5G networks; high-speed industrial ethernet), and blockchain (e.g. supply chain traceability; smart contracts in logistics; license management; and intellectual property) (Agarwal et al., 2022; European Commission, 2020; Ferreira et al., 2023; Zhang & Lu, 2021). Industry 4.0, also referred to as the Fourth Industrial Revolution (4IR), combines these emerging technologies to create smart factories that bring extra efficiency, are more effective, and are increasingly sustainable (Hofmann et al., 2012; Javaid et al., 2022; Tao et al., 2011; Yaqub & Alsabban, 2023). The adoption of digital technologies in manufacturing multinational enterprises (MNEs) has become increasingly vital to remain competitive in the global market, and this adoption has aroused the interest

DOI: 10.1201/9781003461821-7

of professionals in the sector as well as policymakers (Cunha & Maropoulos, 2008; Yang et al., 2021).

At present, the global rate of adoption of digital technologies is still low, with the EU currently far from its goal of 75% adoption (Public First, 2022). The adoption rate of these technologies in MNEs varies significantly across the EU (Brodny & Tutak, 2022). For example, countries such as Germany, Sweden, and Denmark have higher adoption rates of digital technologies in MNEs. In contrast, countries such as Greece and Romania have lower adoption rates of digital technology (Brodny & Tutak, 2022). The differences in adoption rates can be attributed to various factors, including the availability of high-speed infrastructure, government policies, and the level of skilled labor (OECD, 2020). However, the achievement of the Digital Decade objectives by the EU could create economic gains of 2.8 trillion euros, which would represent 21% of the current EU economy (Public First, 2022).

The adoption of digital technologies in MNEs can bring several advantages. AI and the analysis of large volumes of data can help optimize production tactics, reduce waste, and improve product satisfaction by increasing quality (Kockmann et al., 2023). In business systems, the advantages of big data analysis, AI, blockchain, and high-speed infrastructure are recognized (Yu Chung Wang et al., 2022). Big data is a facilitator of Intelligent-Enterprise Resource Planning (ERP) systems, allowing you to streamline business processes and maximize profitability (Chawda & Thakur, 2016) and obtain competitive advantages (Jayawickrama et al., 2019). Robotics and automation can result in greater efficiency and productivity, thereby reducing the burden of employee wages on overall production (Ershova et al., 2023). Digital technologies such as high-speed infrastructures and cloud computing can make it possible to evaluate information in real time, which can facilitate decision-making, especially for more agile and responsive supply chains, allowing enterprise systems to adjust to data streams in real time (Gao et al., 2011). Finally, blockchain can increase transparency and security in supply chains, reducing the possibility of fraud and errors (Pal, 2023) and is a promising ERP transformation technology, enabling the decentralization of work, the traceability of business transactions, and the maintenance of immutable records (Kunduru, 2023).

However, the adoption of digital technologies in multinational manufacturing enterprises also presents very demanding challenges that need to be overcome. One of the main challenges is the cost of implementation and the need for specialized skills and information (Mustapić et al., 2023). In addition, there may be resistance to implementation on the part of employees who are afraid of losing their jobs or who fear that they don't have the necessary training to work with these new technologies (Ferreira et al., 2023). The growing use of digital technologies can also increase issues related to privacy and the protection of information in enterprise systems (Dhirani et al., 2023). These challenges must be overcome in order to ensure the adoption and integration of digital technologies in manufacturing multinationals.

Although there is little literature on the subject, the literature has focused on identifying the factors that drive the adoption of digital technologies, the impact of digital technologies, and the challenges and barriers to the adoption of digital technologies in multinational manufacturing enterprises. Thus, Yang et al. (2021) explored the impact of digitalization on the technological innovations of German small and medium-sized enterprises (SMEs), and technological intelligence and supply chain cooperation were identified as two important factors. Cirillo et al. (2023) identified factors such as cost, compatibility, and complexity as the driving forces behind enterprises' decision to adopt digital technologies. Despite these factors, only 66% of EU manufacturing enterprises say they have adopted at least one digital technology, compared to 78% in the US (Rückert et al., 2020). The impact of digital technologies on manufacturing multinationals in the EU is significant. The implementation of digital technologies could have a substantial impact on economic growth and sustainability in all countries (Bocean & Vărzaru, 2023). The adoption of Industry 4.0 technologies has been associated with back shoring, with Kamp and Gibaja (2021) evaluating this relationship and finding evidence to support it. However, the impact of digitalization on the economy is also influenced by the national economic structure, policies, institutions, and governance (Anderton et al., 2020). Vimal et al. (2023) identified some obstacles to the adoption of technologies in the circular economy, concluding that limited scalability, negative perceptions of technology, and a lack of knowledge and skills are significant obstacles. Along the same lines, Feliciano-Cestero et al. (2023) revealed that many manufacturers face systemic challenges in adopting digital technologies.

However, an important question remains to be answered. RQ1: Can digital technology adoption patterns be identified in EU MNEs? There are not enough studies on the identification of digital technology adoption patterns in multinational manufacturing enterprises (Björkdahl, 2020; Frank et al., 2019; Szalavetz, 2023). The complexity and diversity of these types of enterprise operations and systems can make it difficult to identify clear adoption styles. Therefore, more studies are needed to identify commonalities and variations in the adoption of digital technology in MNEs in order to develop frameworks to understand and analyze these patterns (Axarloglou, 2020; Nambisan & Luo, 2022).

MNEs can have different ownership structures, such as sole ownership by one individual or ownership by more than one individual. The ownership structure can have a huge impact on the adoption of digital technology in organizations. For example, family-owned MNEs may be hesitant to adopt digital technology due to concerns about the costs and complexity of implementation (Atkin et al., 2017). On the other hand, multi-stakeholder MNEs may also have a more positive approach to the adoption of digital technologies, with stakeholders putting forward their views and priorities. While the ownership structure can influence the extent and pattern of the adoption of digital technologies, the activities of MNEs can also have an impact on their adoption (Zhang & Yu, 2023). Therefore, the ownership structure and activities of MNEs can play a key role in determining the level and pattern of digital adoption of this type of enterprise. RQ2: What are the characteristics of MNEs, in terms of propriety, within each pattern of adoption of digital technologies?

This study aims to develop a guide for the adoption of digital technologies by EU MNEs and to analyze the relationship between the ways in which technologies are implemented and the property characteristics of these enterprises. Seven digital technologies were analyzed (Artificial Intelligence, Cloud Computing, Robotics, Smart Devices, Big Data Analytics, High-Speed Infrastructures, and Blockchain). The study is based on a sample of 764 MNEs in the EU.

This study makes four important contributions. First, it has created an empirical contribution to the resource-based view since the adoption of digital technologies (resources) by multinational enterprises in the manufacturing industry can be used as an internal strategy for these enterprises to create competitive advantages. Second, it has contributed to the dynamic capabilities approach, which focuses on the ability of enterprises to proactively adapt to external changes based on the development of resources, particularly technological ones. Third, to our knowledge, this is one of the first studies to analyze the implementation of digital technologies in EU MNEs. Finally, some measures are recommended that can encourage the digitization of MNEs and make them more competitive and able to create or strengthen their competitive advantages: (i) developing digital management skills; (ii) integrating digital transformation into business strategy; (iii) obtaining more accessible and advantageous bank financing, especially when there is a shortage of liquidity and capital in the technology adoption phase; (iv) hiring qualified human resources aligned with the technology to be implemented; (v) reducing barriers to innovation; (vi) facing fewer regulatory hurdles; (vii) facilitating access to data; and (viii) investing a portion of annual turnover in the digitization of enterprises and in R&D. Thus, it is demonstrated how using these technologies in business systems can be advantageous, maximising value creation and distinctive competitive advantages.

7.2 LITERATURE REVIEW

7.2.1 Theoretical Framework

Considering the objective of the study, the theoretical foundation is based on the resource-based view (RBV). For this study, the RBV is deemed to be a theoretical framework that makes it possible to assess the strategic suitability of assets, such as digital technologies, within an enterprise. The RBV approach emphasizes the importance of internal resources and capabilities in achieving competitive advantages rather than external factors such as market conditions and industry trends (Lopes et al., 2021). RBV theory argues that an enterprise's resources and capabilities are the main factors in its performance and that these resources should be unique, valuable, and difficult to imitate (Barney, 1991).

The main assumptions and principles of the theory include the idea that resources can be tangible or intangible and that enterprises should strive to develop and leverage unique resources that are difficult for competitors to imitate (Bharadwaj, 2000). RBV theory also suggests that resources should be valuable in that they allow an enterprise to exploit opportunities or mitigate threats in the environment and that resources should be rare, meaning that they are not widely available in the sector or the market (Lopes et al., 2021). Finally, it argues that resources must be difficult to imitate, which suggests that competitors cannot easily acquire or reproduce them (Barney, 1991).

RBV theory can be applied to the adoption of digital technologies in various ways. For example, the RBV approach can be used to identify and take advantage of unique digital resources and capabilities that are difficult for competitors to replicate (Costa Climent & Haftor, 2021). RBV theory can also be used to assess the value, rarity, and imitability of digital resources, such as software applications and data analysis tools (Hu et al., 2024). In addition, it can help enterprises develop strategies for acquiring or partnering with other organizations to access unique digital resources and capabilities (Lopes et al., 2021). Overall, RBV theory provides a useful framework for understanding the strategic importance of digital resources and capabilities in achieving competitive advantage in today's digital landscape (Rehman Khan et al., 2022).

7.2.2 Definition and Explanation of Digital Technologies

The manufacturing industry in the EU has been adopting digital technologies, including AI, robotics, cloud computing, big data analytics, blockchain, smart devices, and high-speed infrastructures (Ferreira et al., 2023). AI is the simulation of human intelligence processes by machines, namely computer systems (Karinshak, 2024). It is a vast branch of computer science that focuses on building intelligent machines capable of performing tasks that would normally require human intelligence (Ghosh & Thirugnanam, 2021). AI has numerous applications, including natural language processing, speech recognition, decision-making, and problem-solving. AI technology has the potential to revolutionize various sectors, from healthcare to finance and the industrial manufacturing sector. The combination of AI technologies and the Internet of Things (IoT) infrastructure is known as Artificial Intelligence of Things (AIoT) and has the potential to bring about significant changes in the way we live and work (Zhu et al., 2022).

Robotics is the branch of technology that deals with the design, construction, and operation of robots. Robots are machines capable of performing a series of actions autonomously or semi-autonomously (Osawa et al., 2020). Robotics has numerous applications, including in manufacturing, healthcare, and military operations. Robotic technology has the potential to improve efficiency, safety, and productivity in various sectors (Hao et al., 2022; Youn et al., 2020). The Industrial Internet of Things (IIoT) is the use of sensors, actuators, and other smart devices in robotics, such as radio frequency identification tags (Kanagachidambaresan & Bharathi, 2023; Tariq et al., 2022).

Cloud computing refers to the provision of computer services, including servers, storage, databases, networks, software, analytics, and intelligence, over the internet (Ouanouki et al., 2015). This technology allows organizations to access IT resources on demand without the need for a significant initial investment in infrastructure (Al-Azzawi & Kaya, 2021; Avram, 2014). Cloud computing has numerous applications, including data storage, software development, and web hosting. It has the potential to improve efficiency and reduce costs for organizations of all sizes (Xie et al., 2022). Enabling technologies related to Industry 5.0, such as cloud computing, blockchain, big data analytics, IoT, and 6G networks, have the potential to bring about significant changes in the manufacturing industry (Maddikunta et al., 2022; Xiang et al., 2024). An integrated cloud-based solution that harnesses the power of AI, IoT, and advanced analytics is known as IIoT and has the potential to maximize asset performance and extend asset lifecycles (Moustafa et al., 2022; Sakao & Nordholm, 2021).

Big data analytics and blockchain are two technologies that have become increasingly important in recent years. Big data analytics refers to the tools and procedures used to transform large amounts of data into meaningful information. With the growth of the internet and the increasing use of digital devices, the amount of data generated has exploded, making BDA crucial for enterprises looking to gain a competitive advantage (Akter et al., 2023; Ibidun et al., 2023). On the other hand, blockchain is a decentralized and distributed ledger technology that enables secure, transparent, and tamper-proof transactions (Antal et al., 2021; Jena & Dash, 2021). These technologies also show great potential to revolutionize sectors such as finance, healthcare, and supply chain management, improving efficiency, security, and accountability (Benzidia et al., 2023; Fernández-Caramés et al., 2019).

Smart devices are a category of electronic devices that are connected to the internet and can communicate with other devices (Alam & Benaida, 2018). They include smartphones, tablets, smartwatches, and smart home devices such as thermostats, security systems, and lighting (Bugeja et al., 2018; Wang et al., 2023). IIoT is a subset of smart devices that are used in the industrial sector (Oh et al., 2022). Smart devices and IIoT have the potential to improve efficiency and productivity by providing real-time data and enabling automation (Babayigit & Abubaker, 2023).

High-speed infrastructures refer to technologies that enable fast and reliable communication between devices and systems. These infrastructures are crucial for the effective operation of technologies such as AI, cloud computing, and IoT (Low et al., 2011; Wang et al., 2021). With the growing demand for data processing and storage, high-speed infrastructures such as fiber optic networks and 5G networks have become essential (Alsharari et al., 2023). These provide the bandwidth and speed needed to support the growing number of connected devices and the huge amounts of data they generate. With the continued development of high-speed infrastructures, the potential for emerging technologies such as AI and IoT to transform industries and drive innovation is enormous. As such, high-speed infrastructures offer high growth potential in the adoption of digital technologies in MNEs, and their addition has become increasingly frequent in enterprises in recent years (Borangiu et al., 2019; Nadiri et al., 2018). These technologies include AI, cloud computing, robotics, smart devices, big data analytics, high-speed infrastructures, and blockchain. According to Yaqub and Alsabban (2023), these Industry 4.0 digital technologies can have a significant impact on the manufacturing process in MNEs. AI, for example, can be used to optimize production processes, while cloud computing can facilitate collaboration and data sharing between different sites and departments. Robotics can automate repetitive tasks, while big data analytics can be used to identify patterns and trends in production data. Blockchain technology can be utilized to promote device-level data transmission and manufacturing services (Tripathi et al., 2023).

Although the adoption of digital technologies in MNEs can bring significant benefits, such as increased efficiency, productivity, and profitability, there are also challenges to face (Ciulli & Kolk, 2023). One of the main challenges is the need for significant investment in new technologies, including hardware, software, and training. In addition, integrating these technologies into existing production processes can be complex and time-consuming. There may also be concerns about data security, privacy, and intellectual property rights. However, the potential benefits of these technologies often outweigh the challenges, making their adoption a worthwhile investment for many MNEs (Varriale et al., 2023).

More specifically, AI and robotics have been widely adopted in the manufacturing industry, bringing numerous benefits to MNEs. The combination of AI and robotics can improve the efficiency and precision of manufacturing processes, reducing labor costs and increasing productivity (Varriale et al., 2023). AI and robotics can also improve the quality of products and services, as well as enhance safety and security in the workplace. In addition, they can enable multinational enterprises in the manufacturing sector to create new products and services, as well as improve existing ones, by providing real-time data and analysis (Ahi et al., 2022). The integration of AI and robotics into the manufacturing process can lead to significant improvements in the overall performance of MNEs, making them valuable assets.

Cloud computing allows MNEs to store and access data from anywhere, at any time, and on any device (Ooi et al., 2018; Xu, 2012). This technology can strengthen collaboration and communication between teams, as well as improve the speed and efficiency of manufacturing processes. On the other hand, the analysis of large volumes of data can help MNEs make more informed decisions by providing valuable information on customer behavior, market trends, and product performance (Tripathi et al., 2023). This technology can also help manufacturing MNEs optimize their supply chain, reduce waste, and improve sustainability. Blockchain technology, meanwhile, can provide greater security and transparency in the manufacturing process, allowing MNEs to track products and transactions in real time (Tripathi et al., 2023). This technology can also improve supply chain management, reduce fraud and counterfeiting, and increase trust between stakeholders.

Finally, smart devices and high-speed infrastructures are other digital technologies that MNEs can adopt. Smart devices, such as sensors and wearables, can provide real-time data on the manufacturing process, allowing MNEs in the manufacturing sector to optimize their operations and improve product quality (Yaqub & Alsabban, 2023). High-speed infrastructures, such as 5G networks, can enhance communication and collaboration among teams, as well as improve the speed and efficiency of manufacturing processes (Khan et al., 2023). These technologies can also enable MNEs to create new products and services, as well as improve existing ones. Overall, the adoption of smart devices and high-speed infrastructures can lead to significant improvements in the performance and competitiveness of MNEs.

7.2.3 Property Characteristics of Manufacturing Multinational Enterprises

One of the most important issues in the management of MNEs is how they deal with the adoption of digital technologies. As seen above, these technologies can bring benefits such as greater efficiency, quality, innovation, and competitiveness but also challenges such as the need for investment, training, integration, and security (Ciulli & Kolk, 2023; Varriale et al., 2023). In this context, one variable that can influence the decision and performance of MNEs is the property characteristic of the enterprise, that is, whether it is exclusively owned by one person, shared by more than one person, part of a national or international group, or predominantly family-owned.

Exclusive ownership means the ownership of an asset by a single person or entity, which gives them total control and decision-making power over that asset (Chhibber & Majumdar, 1999; Hansmann, 2000). Although the literature does not agree, in MNEs, sole ownership can facilitate the adoption of digital technologies since decision-making processes can be simplified and accelerated. With a clear chain of command and decision-making authority, sole ownership can allow for a more efficient implementation of digital technologies, as there are fewer decision-makers to consult and fewer potential obstacles to adoption. In addition, sole ownership can enable greater investment in digital technologies, as the enterprises' financial resources are concentrated in a single entity (Dawar & Frost, 1999).

Shared ownership, on the other hand, refers to the ownership of assets by more than one person or entity (Hansmann, 2000; Raff et al., 2009). In MNEs, shared ownership can pose challenges to the adoption of digital technologies. Decision-making processes can be more complex and time-consuming, as various stakeholders need to be consulted and a consensus reached before moving forward with the adoption of digital technologies (Liang et al., 2022).

Ownership of national or international assets refers to the ownership of assets that are part of a national or international entity (Kano et al., 2020). In MNEs, national or international ownership can have an impact on the adoption of digital technologies. International enterprises, which are generally larger, find it easier to adopt digital technologies. They can count on more sources of funding, invest a portion of their turnover in digitalization, follow government regulations without major problems, have the technology and qualified personnel, and operate in more stable environments (Roca & O'Sullivan, 2020; Senna et al., 2022; Stornelli et al., 2021). On the other hand, in multinational enterprises, there can be greater bureaucracy and administrative formalities, slowing down decision-making processes and hindering the adoption of digital technologies (Cuervo-Cazurra et al., 2021). These processes could be more streamlined in national enterprises. Ultimately, the impact of national or international ownership on the adoption of new technologies in MNEs may depend on the specific circumstances of each enterprise and the resources and support available to them (Dawar & Frost, 1999).

Family business groups are enterprises that are legally distinct but coordinated to act together (Xavier et al., 2020). These groups are mainly controlled by one family or a small number of individuals (Vecchi et al., 2014). They bring together several enterprises and are often the largest private enterprises in their respective countries. In addition, family businesses are frequently characterized by a strong sense of tradition and an orientation toward long-term goals rather than short-term gains (Swoboda & Olejnik, 2013). This can make it difficult to adopt new technologies and adapt to changing market conditions (Yabushita & Suehiro, 2014). However, family businesses can also be more flexible and agile than other types of enterprises, allowing them to respond quickly to new opportunities (Discua Cruz et al., 2019). Family business groups can provide valuable information about the needs and preferences of local markets, which can serve as a basis for developing and implementing technologies to increase their competitiveness and achieve their strategic objectives (Cuervo-Cazurra, 2006).

7.3 METHODS

7.3.1 Sample

The sample for this study was obtained from the Flash Eurobarometer 486 database – SMEs, Start-ups, Scale-ups, and Entrepreneurship promoted by Gesis – Leibniz Institute for Social Sciences, published in 2020 (European Commission, 2020). This database contains data on enterprises from the 27 EU member states and 12 countries outside the EU (Bosnia and Herzegovina, Kosovo, Canada, Iceland, Japan, North Macedonia, Norway, Serbia, Turkey, UK, USA, and Brazil). The data was collected between February 19, 2020, and May 5, 2020, and the targeted audience of this questionnaire is someone with decision-making responsibilities (managing director, general manager, chief executive officer, financial director), someone leading the commercial activities (commercial manager, sales manager, marketing manager), or a legal officer. This is a non-probabilistic sample. All interviews were carried out via Computer-Assisted Telephone Interviewing. The original sample contained 16,365 enterprises from different sectors of activity. Given the study's objective, the sample was filtered by enterprises in the manufacturing sector in countries belonging to the EU with a multinational presence (activities in more than one country), resulting in a final sample of 764 manufacturing MNEs.

The sample is made up of multinationals from the 27 EU countries (Figure 7.1), with the most significant countries being Italy (6.7%) and Denmark (6.5%).

7.3.2 Measurement of Variables

Regarding the digitalization of manufacturing MNEs, the questionnaire collected responses from the enterprises surveyed regarding: "Which of the following digital technologies has your enterprise adopted so far?" The question allowed multiple answers across seven digital technologies (Table 7.1): i) artificial intelligence,

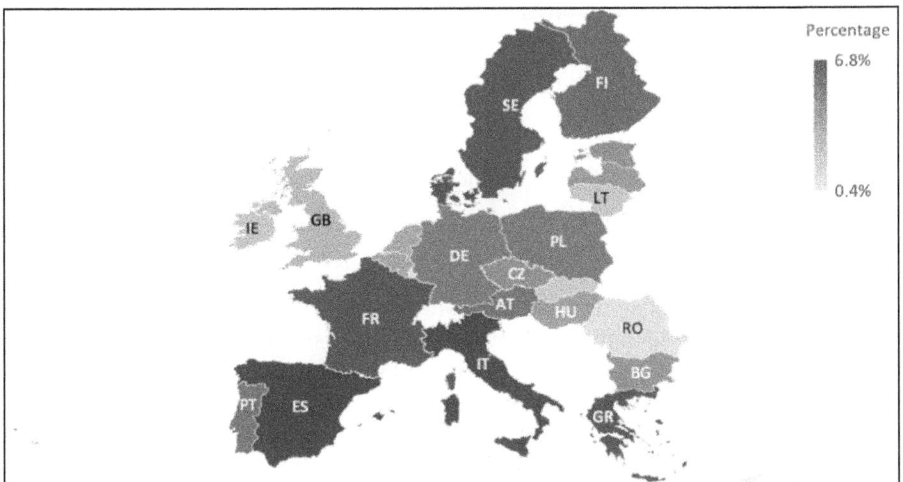

FIGURE 7.1 Geographic dispersion of manufacturing MNEs in the sample.

Note: AT – Austria; BE – Belgium; BG – Bulgaria; CY - Cyprus (Republic); CZ - Czech Republic; DE – Germany; DK – Denmark; EE – Estonia; ES -Spain; FI – Finland; FR – France; GR – Greece; HR – Croatia; HU – Hungary; IE – Ireland; IT – Italy; LT – Lithuania; LU – Luxembourg; LV – Latvia; MT – Malta; NL – The Netherlands; PL – Poland; PT – Portugal; RO – Romania; SE – Sweden; SI – Slovenia; SK – Slovakia.

TABLE 7.1
Digital Technologies

Digital Technology	Description
Artificial intelligence	TEC1 – Artificial intelligence, for example, machine learning or technologies for identifying objects or people
Cloud computing	TEC2 – Cloud computing, that is, storing and processing files or data on remote servers hosted on the internet
Robotics	TEC3 – Robotics, that is, robots used to automate processes such as construction or design
Smart devices	TEC4 – Smart devices, for example, smart sensors and smart thermostats
Big data analytics	TEC5 – Big data analytics, for example, data mining and predictive analysis
High-speed infrastructures	TEC6 – High-speed infrastructures
Blockchain	TEC7 – Application of blockchain technology

ii) cloud computing, iii) robotics, iv) smart devices, v) big data analytics, vi) high-speed infrastructures, and vii) blockchain. This variable has a binary response: 1 – Yes, adopted; 0 – Did not adopt.

Regarding the ownership characteristics of manufacturing MNEs, four variables were considered: i) enterprise size measured by the number of workers (medium entities: 50 to 249 workers; large enterprise: more than 250 workers); ii) year of enterprise foundation (measured by three categories: before 2010; between 2011 and 2014; after 2015); iii) annual turnover in 2019 (measured by six categories: less than 500,000 euros; from 500,000 euros to 1 million euros; from 1 million euros to 5 million euros; from 5 million euros to 10 million euros; from 10 million euros to 50 million euros; more than 50 million euros); iv) ownership (measured by four categories: exclusive property of one person; property of more than one person; part of a national or international and predominantly family business group).

7.3.3 Methodology

In this study, a quantitative methodology was applied. Firstly, a statistical analysis of the variables considered in this study was carried out using the SPSS software (v.25). Then, a correspondence analysis was carried out to relate the adoption of digital technologies implemented by manufacturing MNEs with their characteristics in terms of property. This statistical analysis allows you to explore associations between two variables, similar to principal components analysis (Greenacre & Vrba, 1984).

7.4 RESULTS

7.4.1 Statistical Description of Adoption of Digital Technologies

Table 7.2 shows the frequency of implementation of digital technologies by manufacturing MNEs. The digital technology most adopted by these enterprises is cloud computing (59.9%), and the least adopted is blockchain (6.3%).

TABLE 7.2
Frequency (in %) of Adoption of Digital Technologies

Digital Technology	Frequency (%)
Artificial intelligence	13.2
Cloud computing	59.9
Robotics	36.1
Smart devices	45.8
Big data analytics	22.9
High-speed infrastructures	40.4
Blockchain	6.3

If we consider the average number of digital technologies adopted by manufacturing MNEs by country (Appendix A.1 and Figure 7.1) we see that Luxembourg (M = 4.33), Finland (M = 3.02), and Austria (M = 2.92) are the countries with the highest implementation of digital technologies. Italy (M = 0.98) and Lithuania (0.86) are the countries where manufacturing MNEs show the lowest adoption of digital technologies (Figure 7.2).

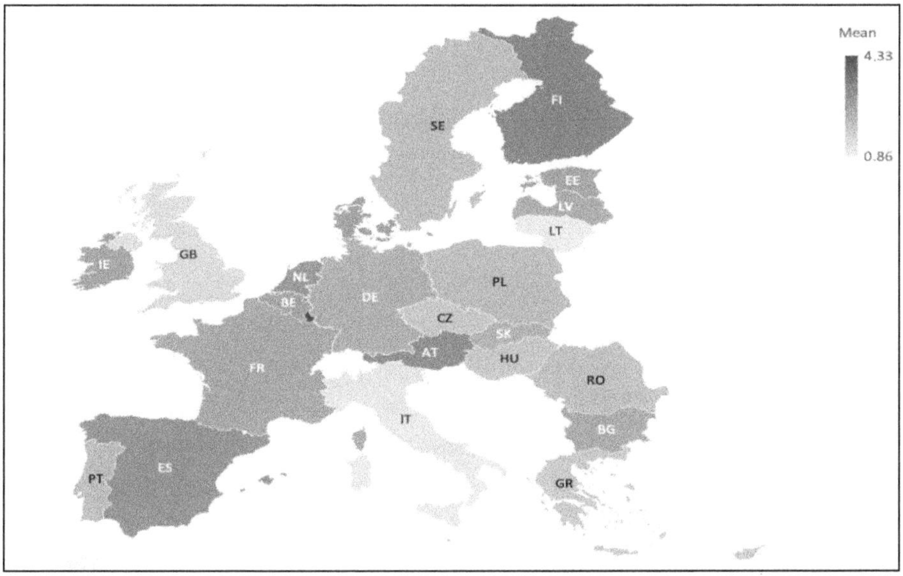

FIGURE 7.2 Average implementation of digital technologies in manufacturing MNEs by countries.

Note: AT – Austria; BE – Belgium; BG – Bulgaria; CY - Cyprus (Republic); CZ - Czech Republic; DE – Germany; DK – Denmark; EE – Estonia; ES -Spain; FI – Finland; FR – France; GR – Greece; HR – Croatia; HU – Hungary; IE – Ireland; IT – Italy; LT – Lithuania; LU – Luxembourg; LV – Latvia; MT – Malta; NL - The Netherlands; PL – Poland; PT – Portugal; RO – Romania; SE – Sweden; SI – Slovenia; SK – Slovakia.

7.4.2 Statistical Description of the Characteristics of Manufacturing MNEs

The sample consists of 764 manufacturing MNEs. The statistical description of enterprise characteristics found in Table 7.3. Regarding size measured by the number of workers, 78% of the enterprises have 250 employees or more. Regarding age, 88.3% of the enterprises were founded before 2014 and only 4.5% after 2018. The turnover of most of these enterprises in 2019 was more than 5 million euros (60.9%). Regarding ownership, 44.2% are owned by more than one person, 17.7% by a single person, 26.8% are business groups, and 28.9% are family-owned.

7.4.3 Relationship between Patterns of Digital Technology Adoption and Properties of Manufacturing MNEs

To analyze the adoption patterns of digital technologies by manufacturing MNEs, we created four patterns: i) low engagement (adoption of zero or one technology); ii) rare engagement (adoption of two or three technologies); iii) moderate engagement

TABLE 7.3
Frequency (in %) of Properties Measures

Variables	Frequency (%)
Employees	
<49	4.9
Between 50 and 249	17.1
>250	78.0
Foundation Year	
<2014	88.3
Between 2015 and 2017	7.2
>2018	4.5
Annual Turnover	
<500,000 euros	11.2
More than 500,000 to 1 million euros	8.8
More than 1 million euros to 5 million euros	19.1
More than 5 million euros to 10 million euros	14.8
More than 10 million euros to 50 million euros	36.0
More than 50 million euros	10.1
Ownership ([a])	
Exclusive property of one person	17.7
Owned by more than one person	44.2
Part of a national or international	26.8
Co-owned by a public entity	2.9
Co-owned by a venture capital firm	3.5
Co-owned by a "business angel"	1.4
Predominantly family business group	28.9
Jointly owned by its members (e.g. cooperative, mutual society)	5.4

Note:
[a] Binary multiple response.

TABLE 7.4
Frequency (in %) of Adoption Patterns of Digital Technologies

Pattern of Adoption of Digital Technology	Frequency (%)
Low engagement	36.8
Rare engagement	41.0
Moderate engagement	18.6
High engagement	3.7

(adoption of four or five technologies); and iv) high engagement (adoption of six or seven technologies). The frequency of these patterns in manufacturing MNEs is shown in Table 7.4. The results show that 41% of these enterprises are in a "rare engagement" pattern; they have adopted two or three digital technologies, while 77.8% of the enterprises implemented a maximum of three digital technologies (low and rare engagement standards).

A correspondence analysis was carried out to relate the adoption patterns of digital technologies with the property characteristics of enterprises. The property characteristics used were: i) enterprise size measured by the number of workers, ii) year of enterprise foundation; and iii) annual turnover in 2019. It was impossible to relate the adoption patterns of digital technologies with enterprise ownership, as this question has a binary answer. Figure 7.3 presents the correspondence analysis results between the characteristics of manufacturing MNEs and the adoption patterns of digital technologies.

Regarding the relationship between the adoption patterns of digital technologies and the size of MNEs, through the correspondence analysis we can observe that enterprises with a low engagement pattern are the enterprises with fewer workers. In contrast, enterprises with a high engagement pattern are the largest (more than 250 employees). Enterprises that were founded several years ago have less engagement in the adoption of technologies. The enterprises with the highest annual turnover (above 50 million euros) show the greatest engagement in adopting digital technologies.

7.5 DISCUSSION AND IMPLICATIONS

7.5.1 Discussion of Results

The increasing globalization of business means that MNEs dominate global economic relations due to their transnational implantation, size, and power (López et al., 2019). These enterprises are a reference in terms of business models to replicate. Within MNEs, manufacturing economic activity stands out due to its geographic location and representativeness (Eurostat, 2022). The implementation of Industry 4.0 in manufacturing enterprises has as its main driver the adoption of digital technologies (Kagermann et al., 2013; Xu et al., 2018). This study aims to evaluate the intensity and forms of adoption of digital technologies by European manufacturing MNEs, as well as their relationship with the characteristics of properties of these enterprises. We contribute with new empirical results by demonstrating the degree

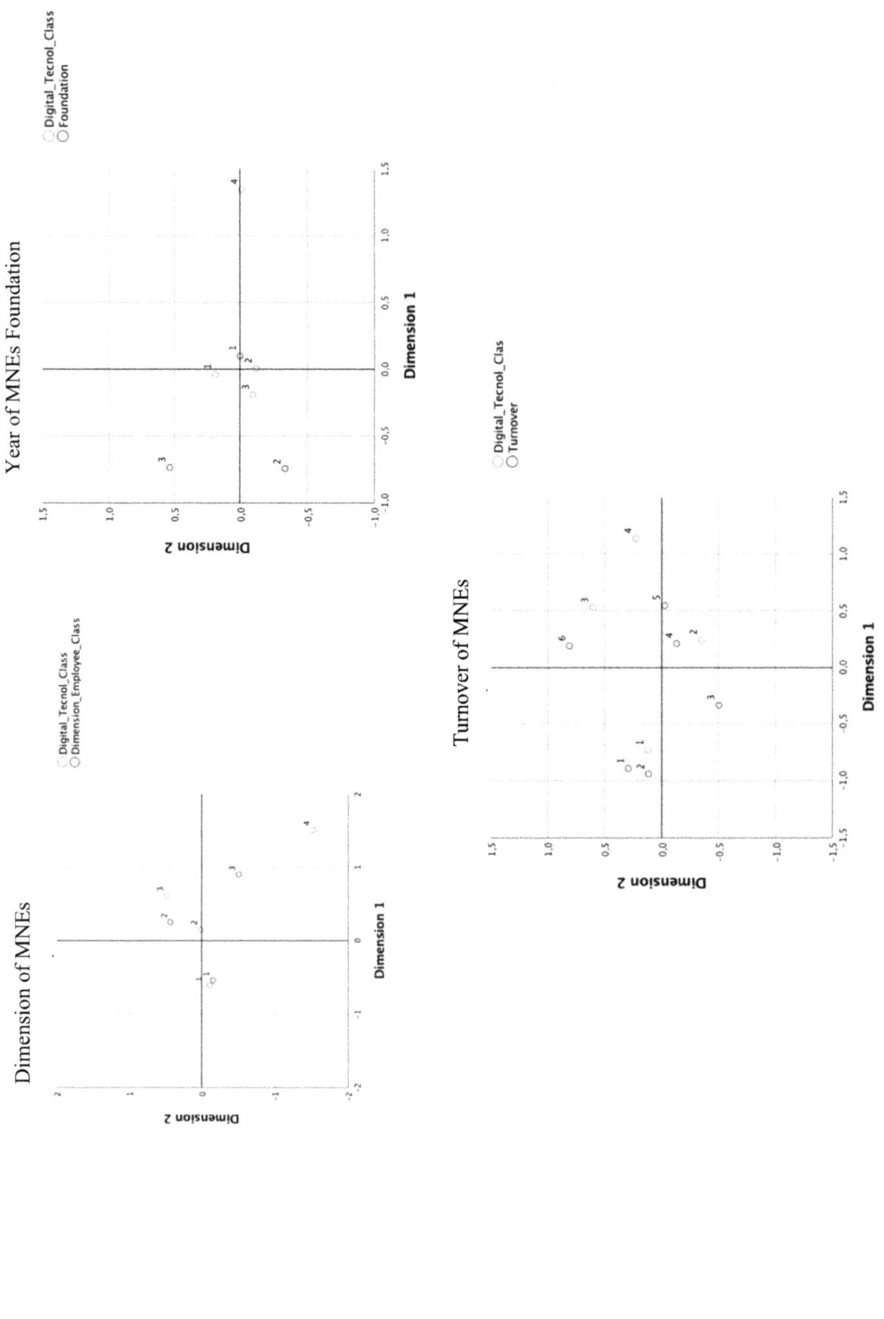

FIGURE 7.3 Symmetric plot of properties of manufacturing MNEs and patterns of digital technology adoption.

of engagement of these enterprises with advanced digital technologies (artificial intelligence, cloud computing, robotics, smart devices, big data analytics, high-speed infrastructures, and blockchain) and by demonstrating that the intensity of technology adoption of digital technologies is related to the property characteristics of manufacturing MNEs.

The results demonstrate that most manufacturing MNEs in the European Union have low engagement in adopting advanced digital technologies, as demonstrated by Li et al. (2024) and Ferreira et al. (2023). Many MNEs implement digital technologies only due to legal requirements and regulatory standards, particularly in transitioning from linear to circular production (Torres de Oliveira et al., 2023). However, the adoption of digital technologies should be the driver of competitive advantages and the processes and internationalization of these enterprises (Bolwijn et al., 2018; Huber et al., 2022; Li et al., 2024) but also environmental, economic, and social sustainability (Jayashree et al., 2022; Müller et al., 2018; Onu & Mbohwa, 2021). Therefore, the adoption of digital technologies must be proactive on the part of manufacturing MNEs until technological capabilities are an internal driver for these enterprises and are incorporated into their business model (Li & Xiong, 2022).

Furthermore, the results demonstrate that the adoption patterns of digital technologies are related to the characteristics of manufacturing MNEs in terms of property. The enterprises with the greatest engagement in adopting the digital technologies considered in this study have the most employees, were founded less than ten years ago, and have a very high annual turnover (more than 50 million euros). The positive relationship between the size (both by number of workers and by annual turnover) of manufacturing MNEs and the adoption of digital technology was found by Buer et al. (2021) and Fu et al. (2021). This relationship can be based on the fact that larger enterprises face fewer obstacles in adopting digital technologies due to the ease of obtaining financing, the possibility of investing part of their turnover in digitalization, having less difficulty in implementing government regulations, possessing technological capabilities and adequate human resources, and experiencing less uncertain environments (Roca & O'Sullivan, 2020; Senna et al., 2022; Stornelli et al., 2021).

7.5.2 Theoretical Implications

This study presents three critical theoretical contributions. First, it contributed empirically to the resource-based view since the adoption of digital technologies (resources) by manufacturing MNEs can be used as an internal strategy for these enterprises to create competitive advantages. This way, MNEs can create resources and capabilities to create sustainable competitive advantages (Roh et al., 2024). Second, it contributed to the dynamic capabilities approach, which focuses on enterprises' ability to proactively adapt to external changes based on the development of resources, particularly technological ones. Dynamic capabilities encourage enterprises to be proactive instead of reactive and to strategically define their long-term sustainable evolution (Bari et al., 2022; Roh et al., 2024). Third, to our knowledge, this is the first study to analyze the implementation of digital technologies by European manufacturing MNEs.

This is important because this study's sample contains data on a large number of manufacturing MNEs from 27 EU countries and the adoption of seven advanced digital technologies. MNEs, due to their implementation and control of global businesses, are a reference case study, allowing us to understand the implementation pattern of digital technologies and relate this pattern to the characteristics of MNE ownership.

7.5.3 Practical Implications

Based on the results of this study, a set of practical implications for manufacturing MNEs and policymakers can be inferred. For MNEs, the results demonstrated that most enterprises have a low implementation of digital technologies. As such, MNEs must strategically define the adoption of more digital technologies to become more competitive and proactively achieve or solidify their competitive advantages. Therefore, manufacturing MNEs must evaluate the advantages of adopting certain digital technologies in their Intelligent-ERP systems. Some drivers can boost the digitalization of MNEs (Stornelli et al., 2021): investing in digital management skills, including digital transformation in the business model; easier and cheaper access to bank financing, particularly with a lack of interest and capital during the technology implementation phase; attracting specialized human resources, suitable for the technology to be adopted; lower barriers to innovation; fewer regulatory obstacles; ease of access to data; and investing part of its turnover annually in the digitalization of enterprises and R&D.

Consequently, governments must take an active role in removing or minimizing MNEs' obstacles in adopting digital technologies in enterprise systems. They must invest in offering specialized training in digital technologies for enterprises' human resources, provide technological monetary vouchers to encourage enterprises to adopt technologies, give a less bureaucratized legal and administrative environment, improve the quality of support services for enterprises, particularly at the level of business associations, invest in information technology infrastructure, such as high-speed internet connection, invest in IT security, and provide tax benefits for the most digitalized enterprises, measured by their contribution to environmental sustainability (Stornelli et al., 2021).

7.6 CONCLUSION

The adoption of advanced digital technologies such as AI, cloud computing, robotics, smart devices, big data analytics, high-speed infrastructures, and blockchain has been recognized as advantageous for enterprise systems due to improving performance, maximizing economic value, and competitive advantages obtained. This study aimed to analyze the implementation of digital technologies in EU manufacturing MNEs and relate the adoption patterns of these technologies to the ownership characteristics of these enterprises. The results made it possible to answer the research questions initially formulated. First, most manufacturing MNEs have low digital technology adoption standards, which are not uniform across EU countries. Later, it was found that the adoption patterns of digital technologies are related to the characteristics of enterprises. Thus, the manufacturing MNEs with more workers, founded more recently (less than ten years), and with very high annual turnover (more than 50 million euros) show greater engagement in adopting digital technologies.

This study has some limitations. The data sample used is secondary, and as such, the variables used in this study were limited to the data collected, namely in terms of enterprises' property characteristics. It would be interesting to include in future studies the relationship between the adoption patterns of digital technologies and the annual turnover investment in R&D and digital technologies, the destination of goods sold (final consumer, other enterprises, or public administration), and the data on supply chain value of these enterprises. In this study, only manufacturing MNEs were considered, but it would be interesting to compare them with MNEs from other sectors of economic activity in the future. Furthermore, it would be interesting to relate the adoption patterns of digital technologies with countries' macroeconomic data, government regulatory standards, levels of compliance with the 2030 Agenda objectives, and the obstacles that manufacturing MNEs face in their digital transformation.

APPENDIX A.1 AVERAGE IMPLEMENTATION OF DIGITAL TECHNOLOGIES IN MANUFACTURING MNES BY COUNTRY

Country	Mean
AT	2.92
BE	2.46
BG	2.26
CY	1.70
CZ	1.63
DE	2.17
DK	2.60
EE	2.45
ES	2.75
FI	3.02
FR	2.34
GB	1.32
GR	1.66
HU	1.67
IE	2.56
IT	0.98
LT	0.86
LU	4.33
LV	2.32
MT	3.00
NL	2.75
PL	1.89
PT	1.93
RO	1.80
SE	1.92
SK	2.08

ACKNOWLEDGMENTS

NECE-UBI, Research Centre for Business Sciences, Research Centre and this work are funded by FCT – Fundação para a Ciência e a Tecnologia, IP, project UIDB/04630/2020.

REFERENCES

Agarwal, P., Swami, S., & Malhotra, S. K. (2022). Artificial Intelligence Adoption in the Post COVID-19 New-Normal and Role of Smart Technologies in Transforming Business: A Review. *Journal of Science and Technology Policy Management*. https://doi.org/10.1108/JSTPM-08-2021-0122

Ahi, A. A., Sinkovics, N., Shildibekov, Y., Sinkovics, R. R., & Mehandjiev, N. (2022). Advanced technologies and international business: A multidisciplinary analysis of the literature. *International Business Review*, *31*(4), 101967. https://doi.org/10.1016/j.ibusrev.2021.101967

Akter, S., Fosso Wamba, S., Sajib, S., Hossain, S., Akter, S., & Fosso Wamba, S. (2023). Chapter 1: Introduction to the Handbook of Big Data Research Methods. In *Handbook of Big Data Research Methods* (pp. 1–10). Edward Elgar Publishing. https://doi.org/10.4337/9781800888555.00005

Alam, T., & Benaida, M. (2018). CICS: Cloud–internet communication security framework for the internet of smart devices. *International Journal of Interactive Mobile Technologies (iJIM)*, *12*(6), 74–84. https://doi.org/10.3991/ijim.v12i6.6776

Al-Azzawi, T., & Kaya, T. (2021). The impact of cloud computing on organizational performance. *International Journal of Cloud Applications and Computing (IJCAC)*, *11*(4), 136–151. https://doi.org/10.4018/IJCAC.2021100108

Alsharari, M., Aliqab, K., Ali, F., & Armghan, A. (2023). Integrated free-space optics and fiber optic network performance enhancement for sustaining 5G high capacity communications. *International Journal of Optics*, *2023*, 8685686. https://doi.org/10.1155/2023/8685686

Anderton, R., Jarvis, V., Labhard, V., Petroulakis, F., Rubene, I., & Vivian, L. (2020). *The Digital Economy and the Euro Area*. European Central Bank. Retrieved January 29, 2024 from https://www.ecb.europa.eu/pub/economic-bulletin/articles/2021/html/ecb.ebart202008_03~da0f5f792a.en.html

Antal, C., Cioara, T., Anghel, I., Antal, M., & Salomie, I. (2021). Distributed ledger technology review and decentralized applications development guidelines. *Future Internet*, *13*(3), 62. https://doi.org/10.3390/fi13030062

Atkin, D., Chaudhry, A., Chaudry, S., Khandelwal, A. K., & Verhoogen, E. (2017). Organizational barriers to technology adoption: Evidence from Soccer-Ball producers in Pakistan. *The Quarterly Journal of Economics*, *132*(3), 1101–1164. https://doi.org/10.1093/qje/qjx010

Avram, M. G. (2014). Advantages and challenges of adopting cloud computing from an enterprise perspective. *Procedia Technology*, *12*, 529–534. https://doi.org/10.1016/j.protcy.2013.12.525

Axarloglou, K. (2020). Digitalization: Disrupting the Business Models of Multinationals. In C. Nikas (Ed.), *Economic Growth in the European Union: Analyzing SME and Investment Policies* (pp. 21–32). Springer International Publishing. https://doi.org/10.1007/978-3-030-48210-7_3

Babayigit, B., & Abubaker, M. (2023). Industrial internet of things: A review of improvements over traditional SCADA systems for industrial automation. *IEEE Systems Journal*, 1–14. https://doi.org/10.1109/jsyst.2023.3270620

Bari, N., Chimhundu, R., & Chan, K.-C. (2022). Dynamic capabilities to achieve corporate sustainability: A roadmap to sustained competitive advantage. *Sustainability*, *14*(3), 1531. https://doi.org/10.3390/su14031531

Barney, J. (1991). Firm resources and sustained competitive advantage. *Journal of Management*, *17*(1), 99–120. https://doi.org/10.1177/014920639101700108

Benzidia, S., Bentahar, O., Husson, J., & Makaoui, N. (2023). Big data analytics capability in healthcare operations and supply chain management: The role of green process innovation. *Annals of Operations Research*. https://doi.org/10.1007/s10479-022-05157-6

Bharadwaj, A. S. (2000). A resource-based perspective on information technology capability and firm performance: An empirical investigation. *MIS Quarterly*, *24*(1), 169–196. https://doi.org/10.2307/3250983

Björkdahl, J. (2020). Strategies for digitalization in manufacturing firms. *California Management Review*, *62*(4), 17–36. https://doi.org/10.1177/0008125620920349

Bocean, C. G., & Vărzaru, A. A. (2023). EU countries' digital transformation, economic performance, and sustainability analysis. *Humanities and Social Sciences Communications*, *10*(1), 875. https://doi.org/10.1057/s41599-023-02415-1

Bolwijn, R., Casella, B., & Zhan, J. (2018). International Production and the Digital Economy. In R. van Tulder, A. Verbeke, & L. Piscitello (Eds.), *International Business in the Information and Digital Age* (Vol. 13, pp. 39–64). Emerald Publishing Limited. https://doi.org/10.1108/S1745-886220180000013003

Borangiu, T., Trentesaux, D., Thomas, A., Leitão, P., & Barata, J. (2019). Digital transformation of manufacturing through cloud services and resource virtualization. *Computers in Industry*, *108*, 150–162. https://doi.org/10.1016/j.compind.2019.01.006

Brodny, J., & Tutak, M. (2022). Analyzing the level of digitalization among the enterprises of the European Union member states and their impact on economic growth. *Journal of Open Innovation: Technology, Market, and Complexity*, *8*(2), 70. https://doi.org/10.3390/joitmc8020070

Buer, S.-V., Strandhagen, J. W., Semini, M., & Strandhagen, J. O. (2021). The digitalization of manufacturing: Investigating the impact of production environment and company size. *Journal of Manufacturing Technology Management*, *32*(3), 621–645. https://doi.org/10.1108/JMTM-05-2019-0174

Bugeja, J., Jacobsson, A., & Davidsson, P. (2018). Smart connected homes. In *Internet of Things A to Z* (pp. 359–384). https://doi.org/10.1002/9781119456735.ch13

Chawda, R. K., & Thakur, G. (2016, March). Big data and advanced analytics tools. In *2016 symposium on colossal data analysis and networking (CDAN)* (pp. 1–8). IEEE.

Chhibber, Pradeep K., & Majumdar, Sumit K. (1999). Foreign ownership and profitability: Property rights, control, and the performance of firms in Indian industry. *The Journal of Law and Economics*, *42*(1), 209–238. https://doi.org/10.1086/467423

Cirillo, V., Fanti, L., Mina, A., & Ricci, A. (2023). The adoption of digital technologies: Investment, skills, work organisation. *Structural Change and Economic Dynamics*, *66*, 89–105. https://doi.org/10.1016/j.strueco.2023.04.011

Ciulli, F., & Kolk, A. (2023). International business, digital technologies and sustainable development: Connecting the dots. *Journal of World Business*, *58*(4), 101445. https://doi.org/10.1016/j.jwb.2023.101445

Costa Climent, R., & Haftor, D. M. (2021). Business model theory-based prediction of digital technology use: An empirical assessment. *Technological Forecasting and Social Change*, *173*, 121174. https://doi.org/10.1016/j.techfore.2021.121174

Cuervo-Cazurra, A. (2006). Business groups and their types. *Asia Pacific Journal of Management*, *23*(4), 419–437. https://doi.org/10.1007/s10490-006-9012-5

Cuervo-Cazurra, A., Dieleman, M., Hirsch, P., Rodrigues, S. B., & Zyglidopoulos, S. (2021). Multinationals' misbehavior. *Journal of World Business*, *56*(5), 101244. https://doi.org/10.1016/j.jwb.2021.101244

Cunha, P. F., & Maropoulos, P. G. (2008). Digital enterprise technology: Perspectives and future challenges. *International Journal of Computer Integrated Manufacturing*, *21*(2), 125–126. https://doi.org/10.1080/09511920801936297

Dawar, N., & Frost, T. (1999). Competing with giants: Survival strategies for local companies in emerging markets. *Harvard Business Review*, *77*, 119–132.

Dhirani, L. L., Mukhtiar, N., Chowdhry, B. S., & Newe, T. (2023). Ethical dilemmas and privacy issues in emerging technologies: A review. *Sensors*, *23*(3), 1151. https://doi.org/10.3390/s23031151

Discua Cruz, A., Basco, R., Parada, M. J., Malfense Fierro, A., & Alvarado-Alvarez, C. (2019). Resilience and family business groups in unstable economies. In M. Rautiainen, P. Rosa, T. Pihkala, M. J. Parada, & A. D. Cruz (Eds.), *The Family Business Group Phenomenon: Emergence and Complexities* (pp. 315–352). Springer International Publishing. https://doi.org/10.1007/978-3-319-98542-8_13

Ershova, A. S., Gugutishvili, D. M., Lepekhin, A. A., & Tick, A. (2023). Application of robotic process automation technology for business processes in the field of finance and accounting. In I. Ilin, M. M. Petrova, & T. Kudryavtseva, *Digital Transformation on Manufacturing, Infrastructure & Service*.

European Commission. (2020). *Flash Eurobarometer 486: SMEs, Start-ups, Scale-ups and Entrepreneurship*. Publications Office of the European Union. Retrieved January 20, 2024 from https://data.europa.eu/data/datasets/s2244_486_eng?locale=en

Eurostat. (2022). *Structure of Multinational Enterprise Groups in the EU*. European Commission. Retrieved January 20, 2024 from https://ec.europa.eu/eurostat/statistics-explained/index.php?title=Structure_of_multinational_enterprise_groups_in_the_EU#Multinational_enterprise_groups_in_the_EU_and_EFTA_are_mainly_active_in_the_manufacturing_sector.2C_and_employment_is_concentrated_in_a_small_number_of_sectors

Feliciano-Cestero, M. M., Ameen, N., Kotabe, M., Paul, J., & Signoret, M. (2023). Is digital transformation threatened? A systematic literature review of the factors influencing firms' digital transformation and internationalization. *Journal of Business Research*, *157*, 113546. https://doi.org/10.1016/j.jbusres.2022.113546

Fernández-Caramés, T. M., Blanco-Novoa, O., Froiz-Míguez, I., & Fraga-Lamas, P. (2019). Towards an autonomous Industry 4.0 warehouse: A UAV and blockchain-based system for inventory and traceability applications in big data-driven supply chain management. *Sensors*, *19*(10), 2394. https://doi.org/10.3390/s19102394

Ferreira, J. J., Lopes, J. M., Gomes, S., & Rammal, H. G. (2023). Industry 4.0 implementation: Environmental and social sustainability in manufacturing multinational enterprises. *Journal of Cleaner Production*, *404*, 136841. https://doi.org/10.1016/j.jclepro.2023.136841

Frank, A. G., Dalenogare, L. S., & Ayala, N. F. (2019). Industry 4.0 technologies: Implementation patterns in manufacturing companies. *International Journal of Production Economics*, *210*, 15–26. https://doi.org/10.1016/j.ijpe.2019.01.004

Fu, X., Emes, D., & Hou, J. (2021). Multinational enterprises and structural transformation in emerging and developing countries: A survey of the literature. *International Business Review*, *30*(2), 101801. https://doi.org/10.1016/j.ibusrev.2021.101801

Gao, J., Bai, X., & Tsai, W. T. (2011). Cloud testing-issues, challenges, needs and practice. *Software Engineering: An International Journal*, 1(1), 9–23.

Ghosh, M., & Thirugnanam, A. (2021). Introduction to artificial intelligence. In K. G. Srinivasa, G. M. Siddesh, & S. R. M. Sekhar (Eds.), *Artificial Intelligence for Information Management: A Healthcare Perspective* (pp. 23–44). Springer. https://doi.org/10.1007/978-981-16-0415-7_2

Greenacre, M. J., & Vrba, E. S. (1984). Graphical display and interpretation of antelope census data in African wildlife areas, using correspondence analysis. *Ecology*, *65*(3), 984–997. https://doi.org/10.2307/1938070

Hansmann, H. (2000). *The Ownership of Enterprise*. Harvard University Press.

Hao, Y., Zhang, S., Fang, B., Sun, F., Liu, H., & Li, H. (2022). A review of smart materials for the boost of soft actuators, soft sensors, and robotics applications. *Chinese Journal of Mechanical Engineering*, *35*(1), 37. https://doi.org/10.1186/s10033-022-00707-2

Hofmann, D., Margull, R., Dittrich, P. G., & Düntsch, E. (2012). *Smartphone Green Vision at Dawn of Industry 4.0*. Advanced Materials Research.

Hu, Y., Pan, Y., Yu, M., & Chen, P. (2024). Navigating digital transformation and knowledge structures: Insights for small and medium-sized enterprises. *Journal of the Knowledge Economy*. https://doi.org/10.1007/s13132-024-01754-x

Huber, R., Oberlander, A. M., Faisst, U., & Roglinger, M. (2022). Disentangling capabilities for Industry 4.0 – An information systems capability perspective. *Information Systems Frontiers*, 1–29. https://doi.org/10.1007/s10796-022-10260-x

Ibidun, A., Egbuta, O. U., & Akinlabi, B. H. (2023). Big data analytics and competitive advantage: Evidences from healthcare services organisations in Lagos state, Nigeria. *Journal of Information and Technology*, *7*(1), 22–46. https://doi.org/10.53819/81018102t4149

Javaid, M., Haleem, A., Singh, R. P., Suman, R., & Gonzalez, E. S. (2022). Understanding the adoption of Industry 4.0 technologies in improving environmental sustainability. *Sustainable Operations and Computers*, *3*, 203–217. https://doi.org/10.1016/j.susoc.2022.01.008

Jayashree, S., Reza, M. N. H., Malarvizhi, C. A. N., Gunasekaran, A., & Rauf, M. A. (2022). Testing an adoption model for Industry 4.0 and sustainability: A Malaysian scenario. *Sustainable Production and Consumption*, *31*, 313–330. https://doi.org/10.1016/j.spc.2022.02.015

Jayawickrama, U., Liu, S., Hudson Smith, M., Akhtar, P., & Al Bashir, M. (2019). Knowledge retention in ERP implementations: The context of UK SMEs. *Production Planning & Control*, *30*(10–12), 1032–1047. https://doi.org/10.1080/09537287.2019.1582107

Jena, A. K., & Dash, S. P. (2021). Blockchain technology: Introduction, applications, challenges. In S. K. Panda, A. K. Jena, S. K. Swain, & S. C. Satapathy (Eds.), *Blockchain Technology: Applications and Challenges* (pp. 1–11). Springer International Publishing. https://doi.org/10.1007/978-3-030-69395-4_1

Kagermann, H., Wahlster, W., & Helbig, J. (2013). Recommendations for implementing the strategic initiative INDUSTRIE 4.0. Final report of the Industrie, Issue. N. A. o. S. a. Engineering. https://en.acatech.de/wp-content/uploads/sites/6/2018/03/Final_report__Industrie_4.0_accessible.pdf

Kamp, B., & Gibaja, J. J. (2021). Adoption of digital technologies and backshoring decisions: is there a link?. *Operations Management Research*, 14(3), 380–402.

Kanagachidambaresan, G., & Bharathi, N. (2023). Sensors for the industrial internet of things. In *Sensors and Protocols for Industry 4.0: Industrial Applications of Maker Tech* (pp. 29–72). Apress. https://doi.org/10.1007/978-1-4842-9007-1_2

Kano, L., Tsang, E. W. K., & Yeung, H. W. C. (2020). Global value chains: A review of the multi-disciplinary literature. *Journal of International Business Studies*, *51*(4), 577–622. https://doi.org/10.1057/s41267-020-00304-2

Karinshak, E. (2024). Simulations in human-AI crisis management systems. *Journal of ContingenciesandCrisisManagement*,*32*(1),e12525.https://doi.org/10.1111/1468-5973.12525

Khan, A. A., Laghari, A. A., Li, P., Dootio, M. A., & Karim, S. (2023). The collaborative role of blockchain, artificial intelligence, and industrial internet of things in digitalization of small and medium-size enterprises. *Scientific Reports*, *13*(1), 1656. https://doi.org/10.1038/s41598-023-28707-9

Kockmann, N., Schindler, T., & Urbas, L. (2023). AI in Process Industries – Incubator Labs and Use Cases. *Chemie Ingenieur Technik*, *95*(7), 963–963. https://doi.org/10.1002/cite.202370702

Kunduru, A. R. (2023). Blockchain technology for ERP systems: A review. *American Journal of Engineering, Mechanics and Architecture*, *1*(7), 56–63. https://grnjournal.us/index.php/AJEMA/article/view/775

Li, F., Chen, Y., Ortiz, J., & Wei, M. (2024). The theory of multinational enterprises in the digital era: State-of-the-art and research priorities. *International Journal of Emerging Markets*, *19*(2), 390–411. https://doi.org/10.1108/IJOEM-03-2021-0366

Li, K.-S., & Xiong, Y.-Q. (2022). Host country's environmental uncertainty, technological capability, and foreign market entry mode: Evidence from high-end equipment manufacturing MNEs in emerging markets. *International Business Review*, *31*(1), 101900. https://doi.org/10.1016/j.ibusrev.2021.101900

Liang, H., Kou, G., Dong, Y., Chiclana, F., & Herrera-Viedma, E. (2022). Consensus reaching with minimum cost of informed individuals and time constraints in large-scale group decision-making. *IEEE Transactions on Fuzzy Systems*, *30*(11), 4991–5004. https://doi.org/10.1109/TFUZZ.2022.3165373

Lopes, J., Ferreira, J. J., & Farinha, L. (2021). Entrepreneurship and the resource-based view: What is the linkage? A bibliometric approach. *International Journal of Entrepreneurial Venturing*, *13*(2), 137–164. https://doi.org/10.1504/IJEV.2021.114408

López, L.-A., Cadarso, M.-Á., Zafrilla, J., & Arce, G. (2019). The carbon footprint of the U.S. multinationals' foreign affiliates. *Nature Communications*, *10*(1), 1672. https://doi.org/10.1038/s41467-019-09473-7

Low, C., Chen, Y., & Wu, M. (2011). Understanding the determinants of cloud computing adoption. *Industrial Management & Data Systems*, *111*(7), 1006–1023. https://doi.org/10.1108/02635571111161262

Maddikunta, P. K. R., Pham, Q.-V., Prabadevi, B., Deepa, N., Dev, K., Gadekallu, T. R., … Liyanage, M. (2022). Industry 5.0: A survey on enabling technologies and potential applications. *Journal of Industrial Information Integration*, *26*, 100257. https://doi.org/10.1016/j.jii.2021.100257

Moustafa, N., Choo, K. K. R., & Abu-Mahfouz, A. M. (2022). Guest editorial: AI-enabled threat intelligence and hunting microservices for distributed industrial IoT system. *IEEE Transactions on Industrial Informatics*, *18*(3), 1892–1895. https://doi.org/10.1109/TII.2021.3111028

Müller, J. M., Kiel, D., & Voigt, K.-I. (2018). What drives the implementation of Industry 4.0? The role of opportunities and challenges in the context of sustainability. *Sustainability*, *10*(1), 247. https://doi.org/10.3390/su10010247

Mustapić, M., Trstenjak, M., Gregurić, P., & Opetuk, T. (2023). Implementation and use of digital, green and sustainable technologies in internal and external transport of manufacturing companies. *Sustainability*, *15*(12), 9557. https://doi.org/10.3390/su15129557

Nadiri, M. I., Nandi, B., & Akoz, K. K. (2018). Impact of modern communication infrastructure on productivity, production structure and factor demands of US industries: Impact revisited. *Telecommunications Policy*, *42*(6), 433–451. https://doi.org/10.1016/j.telpol.2018.03.008

Nambisan, S., & Luo, Y. (2022). *The Digital Multinational: Navigating the New Normal in Global Business*. The MIT Press. https://doi.org/10.7551/mitpress/13579.001.0001

OECD. (2020). *The Digitalisation of Science, Technology and Innovation*. OECD Publishing. https://doi.org/10.1787/b9e4a2c0-en

Oh, S., Kim, D., Lee, C., & Jeong, J. (2022). Edge-cloud alarm level of heterogeneous IIoT devices based on knowledge distillation in smart manufacturing. *Electronics*, *11*(6), 899. https://doi.org/10.3390/electronics11060899

Onu, P., & Mbohwa, C. (2021). Industry 4.0 opportunities in manufacturing SMEs: Sustainability outlook. *Materials Today: Proceedings*, *44*, 1925–1930. https://doi.org/10.1016/j.matpr.2020.12.095

Ooi, K.-B., Lee, V.-H., Tan, G. W.-H., Hew, T.-S., & Hew, J.-J. (2018). Cloud computing in manufacturing: The next industrial revolution in Malaysia? *Expert Systems with Applications*, *93*, 376–394. https://doi.org/10.1016/j.eswa.2017.10.009

Osawa, M., Okuoka, K., Takimoto, Y., & Imai, M. (2020). Is automation appropriate? Semi-autonomous telepresence architecture focusing on voluntary and involuntary movements. *International Journal of Social Robotics*, *12*(5), 1119–1134. https://doi.org/10.1007/s12369-020-00620-5

Ouanouki, R., Morales, A. G., & April, A. (2015). Should the cloud computing definition include a big data perspective? In D. B. A. M. Khosrow-Pour (Ed.), *Encyclopedia of Information Science and Technology*, 3rd Edition (pp. 1088–1095). IGI Global. https://doi.org/10.4018/978-1-4666-5888-2.ch104

Pal, K. (2023). Managing supply chain digitalization with blockchain technology. In T. Najar, Y. Najar, & A. Aloui (Eds.), *Handbook of Research on Blockchain Technology and the Digitalization of the Supply Chain* (pp. 137–165). IGI Global. https://doi.org/10.4018/978-1-6684-7455-6.ch007

Public First. (2022). *Unlocking Europe's Digital Potential*. Public First. https://awsdigitaldecade.publicfirst.co.uk/

Raff, H., Ryan, M., & Stähler, F. (2009). Whole vs. shared ownership of foreign affiliates. *International Journal of Industrial Organization*, *27*(5), 572–581. https://doi.org/10.1016/j.ijindorg.2009.01.003

Rehman Khan, S. A., Ahmad, Z., Sheikh, A. A., & Yu, Z. (2022). Digital transformation, smart technologies, and eco-innovation are paving the way toward sustainable supply chain performance. *Science Progress*, *105*(4), 00368504221145648. https://doi.org/10.1177/00368504221145648

Roca, J., & O'Sullivan, E. (2020). Seeking coherence between barriers to manufacturing technology adoption and innovation policy. *International Journal of Production Economics*, *230*, 107818. https://doi.org/10.1016/j.ijpe.2020.107818

Roh, T., Xiao, S., & Park, B. I. (2024). MNEs' capabilities and their sustainable business in emerging markets: Evidence from MNE subsidiaries in China. *Journal of International Management*, *30*(1), 101097. https://doi.org/10.1016/j.intman.2023.101097

Rückert, D., Weiss, W., & Revoltella, D. (2020). *Adoption of Digital Technologies by Firms in Europe and the US: Evidence from the EIB Investment Survey*. Centre for Economic Policy Research (CEPR). Retrieved January 29, 2024 from https://cepr.org/voxeu/columns/adoption-digital-technologies-firms-europe-and-us-evidence-eib-investment-survey

Sakao, T., & Nordholm, A. K. (2021). Requirements for a product lifecycle management system using internet of things and big data analytics for product-as-a-service [mini review]. *Frontiers in Sustainability*, *2*, 1–7. https://doi.org/10.3389/frsus.2021.735550

Senna, P. P., Ferreira, L. M. D. F., Barros, A. C., Bonnín Roca, J., & Magalhães, V. (2022). Prioritizing barriers for the adoption of Industry 4.0 technologies. *Computers & Industrial Engineering*, *171*, 108428. https://doi.org/10.1016/j.cie.2022.108428

Stornelli, A., Ozcan, S., & Simms, C. (2021). Advanced manufacturing technology adoption and innovation: A systematic literature review on barriers, enablers, and innovation types. *Research Policy*, *50*(6), 104229. https://doi.org/10.1016/j.respol.2021.104229

Swoboda, B., & Olejnik, E. (2013). A taxonomy of small- and medium-sized international family firms. *Journal of International Entrepreneurship*, *11*(2), 130–157. https://doi.org/10.1007/s10843-012-0101-x

Szalavetz, A. (2023). Digital technologies shaping the nature and routine intensity of shopfloor work. *Competition & Change*, *27*(2), 277–301. https://doi.org/10.1177/10245294221107489

Tao, F., Zhang, L., Venkatesh, V. C., Luo, Y., & Cheng, Y. (2011). Cloud manufacturing: A computing and service-oriented manufacturing model. *Proceedings of the Institution of Mechanical Engineers, Part B: Journal of Engineering Manufacture*, *225*(10), 1969–1976. https://doi.org/10.1177/0954405411405575

Tariq, U., Ahanger, T. A., Ibrahim, A., & Bouteraa, Y. S. (2022). The industrial internet of things (IIoT): An anomaly identification and countermeasure method. *Journal of Circuits, Systems and Computers*, *31*(12), 2250219. https://doi.org/10.1142/s021812662250219x

Torres de Oliveira, R., Ghobakhloo, M., & Figueira, S. (2023). Industry 4.0 towards social and environmental sustainability in multinationals: Enabling circular economy, organizational social practices, and corporate purpose. *Journal of Cleaner Production*, *430*, 139712. https://doi.org/10.1016/j.jclepro.2023.139712

Tripathi, G., Ahad, M. A., & Casalino, G. (2023). A comprehensive review of blockchain technology: Underlying principles and historical background with future challenges. *Decision Analytics Journal*, *9*, 100344. https://doi.org/10.1016/j.dajour.2023.100344

Varriale, V., Cammarano, A., Michelino, F., & Caputo, M. (2023). Critical analysis of the impact of artificial intelligence integration with cutting-edge technologies for production systems. *Journal of Intelligent Manufacturing*. https://doi.org/10.1007/s10845-023-02244-8

Vecchi, A., Piana, B. D., & Cacia, C. (2014). Mapping family business groups from a cross-cultural perspective. In L. Brennan (Ed.), *Enacting Globalization: Multidisciplinary Perspectives on International Integration* (pp. 194–204). Palgrave Macmillan. https://doi.org/10.1057/9781137361943_18

Vimal, K., Sivakumar, K., Kandasamy, J., Venkat, V., & Mani, R. S. (2023). Barriers to the adoption of digital technologies in a functional circular economy network. *Operations Management Research*, *16*(3), 1541–1561. https://doi.org/10.1007/s12063-023-00375-y

Wang, W., Sun, K., Zeng, C., Chen, C., Qiu, W., You, S., & Liu, Y. (2021). Information and communication infrastructures in modern wide-area systems. In H. Haes Alhelou, A. Y. Abdelaziz, & P. Siano (Eds.), *Wide Area Power Systems Stability, Protection, and Security* (pp. 71–104). Springer International Publishing. https://doi.org/10.1007/978-3-030-54275-7_3

Wang, Y., Zhang, X., & Hu, H. (2023). Continuous user authentication on multiple smart devices. *Information*, *14*(5), 274. https://doi.org/10.3390/info14050274

Xavier, W., Camilo, S. P. O., Marcon, R., & Greene, F. (2020). Ownership structure of family business groups. *Revista Visão: Gestão Organizacional*, *9*(2), 240–253. https://doi.org/10.33362/visao.v9i2.2470

Xiang, W., Yu, K., Han, F., Fang, L., He, D., & Han, Q. L. (2024). Advanced manufacturing in Industry 5.0: A survey of key enabling technologies and future trends. *IEEE Transactions on Industrial Informatics*, *20*(2), 1055–1068. https://doi.org/10.1109/TII.2023.3274224

Xie, G., Wei, Y., Le, Y., & Li, R. (2022). Redundancy minimization and cost reduction for workflows with reliability requirements in cloud-based services. *IEEE Transactions on Cloud Computing*, *10*(1), 633–647. https://doi.org/10.1109/TCC.2019.2937933

Xu, L. D., Xu, E. L., & Li, L. (2018). Industry 4.0: State of the art and future trends. *International Journal of Production Research*, *56*(8), 2941–2962. https://doi.org/10.1080/00207543.2018.1444806

Xu, X. (2012). From cloud computing to cloud manufacturing. *Robotics and Computer-Integrated Manufacturing*, *28*(1), 75–86. https://doi.org/10.1016/j.rcim.2011.07.002

Yabushita, N. W., & Suehiro, A. (2014). Family business groups in Thailand: Coping with management critical points. *Asia Pacific Journal of Management*, *31*(4), 997–1018. https://doi.org/10.1007/s10490-014-9388-6

Yang, M., Fu, M., & Zhang, Z. (2021). The adoption of digital technologies in supply chains: Drivers, process and impact. *Technological Forecasting and Social Change*, *169*, 120795. https://doi.org/10.1016/j.techfore.2021.120795

Yaqub, M. Z., & Alsabban, A. (2023). Industry-4.0-enabled digital transformation: Prospects, instruments, challenges, and implications for business strategies. *Sustainability*, *15*(11), 8553. https://doi.org/10.3390/su15118553

Youn, J.-H., Jeong, S. M., Hwang, G., Kim, H., Hyeon, K., Park, J., & Kyung, K.-U. (2020). Dielectric elastomer actuator for soft robotics applications and challenges. *Applied Sciences*, *10*(2), 640. https://doi.org/10.3390/app10020640

Yu Chung Wang, W., Pauleen, D., & Taskin, N. (2022). Enterprise systems, emerging technologies, and the data-driven knowledge organisation. *Knowledge Management Research & Practice*, *20*(1), 1–13. https://doi.org/10.1080/14778238.2022.2039571

Zhang, C., & Lu, Y. (2021). Study on artificial intelligence: The state of the art and future prospects. *Journal of Industrial Information Integration*, *23*, 100224. https://doi.org/10.1016/j.jii.2021.100224

Zhang, X., & Yu, B. (2023). The impact of ownership structure on technological innovation and energy intensity: Evidence from China. *Sustainability*, *15*(11), 8512. https://doi.org/10.3390/su15118512

Zhu, S., Ota, K., & Dong, M. X. (2022). Energy-efficient artificial intelligence of things with intelligent edge. *IEEE Internet of Things Journal*, *9*(10), 7525–7532. https://doi.org/10.1109/jiot.2022.3143722

8 A Framework for Enabling Effective Digital Transformation in Organizations Based in Business Processes

Francisco J. Duarte
University of Minho, Guimarães, Portugal

8.1 INTRODUCTION

The contemporary business landscape is characterized by a digital foundation for most organizational operations. This digital foundation necessitates real-time interoperability with suppliers, partner organizations, and customers. To achieve business value creation, seamless interaction is required between entities in the physical world, such as sensors, actuators, machines, personnel, and products, and digital entities, such as data and algorithms. Business Process Management (BPM) offers a comprehensive framework for orchestrating business activities in this context (Smith & Fingar, 2002), including the ability to improve business processes (Denner et al., 2018) or the ability to continuously evolve and incorporate new capabilities (Kerpedzhiev et al., 2021). Imgrund et al. (2018) posit that BPM initiatives are not the sole preserve of large corporations but can be effectively implemented by Small and Medium-Sized Enterprises (SMEs) to achieve digital transformation objectives. Across diverse organizations, regardless of their nature, size, or industry sector, organizations perform business processes, albeit with varying degrees of explicit management. Widely used standards and business excellence models, such as ISO 9001 (2015) or the EFQM Model (2021), assume and promote the explicit management of business processes. The absence of such explicit management can, for example, foster a huge dispersion in the descriptions of the business activities, leading to internal discussions and knowledge gaps, because processes are described in many places, such as quality standards, worksheets, presentation slides, or even not described at all. In such a way, when an enterprise-wide software system needs to be developed or customized, the requirements collection phase takes more time and is more prone to errors and omissions. For all the presented reasons, business process models are mandatory in the development and evolution of software enterprise-wide systems.

 DOI: 10.1201/9781003461821-8

Porter (1980) states that to win a competitive advantage in any market, a firm needs to be able to deliver a given set of customer benefits at lower costs than competitors or provide customers with a bundle of benefits its rivals cannot match. To achieve these goals, organizations need to focus their activities on customers and need to increase their ability to innovate continuously. This demanding and changing environment drives the call for reorganizations of the internal structure to keep pace with the changing external factors. Hamel (2000) also points out that the fundamental challenge that organizations face is reinventing themselves and their industries not just in times of crisis but continually. Furthermore, the focus on customer requirements drives the need to change internal organizational structures. This involves strategic realignment and coordination of activities to ensure responsiveness to customer needs. Consequently, the use of well-defined and computationally compatible business process models becomes imperative to construct software-based enterprise systems.

Based on strategy, and on the identified strategic success factors, each organization needs to draw a business model. This model is a conceptual tool that contains a big set of elements and their relationships and allows expressing the business logic of a specific organization (Osterwalder et al., 2005) and also contains the architecture of the organization and its network of partners. Thus, managers need to conceive the internal structure of their organizations in such a way to provide value to customers and describe it using standard notations, such as the Business Process Model and Notation (OMG, 2011) or UML Activity Diagrams (OMG, 2017). Due to the need to be aware of the business environment, the business model should be able to support changes, namely those on the internal structure of the organization. Kotarba (2018) states that the digital transformation of an organization has impacts on its business model.

Porter and Millar (1985) mention that information gives organizations competitive advantages. Information management and related technologies have strategic significance for organizations from allowing computer-aided design in technology development to incorporating automation in warehouses (Porter & Millar, 1999). Thus, when an organization reshapes itself to deliver the best value to customers, a proper way to accomplish it is to look for value-added activities, as in a value chain analysis (see Figure 8.1).

According to Porter (1985), a value chain is a chain of activities. Products or services pass through activities to gain value. The value chain categorizes the generic value-adding activities of an organization. Currently, value chains are mature and key models for the digital economy (Miao, 2021) or for restructuring in big corporations (López et al., 2022), and can map and organize activities in all areas of an organization, such as in human resources and operations.

The explicit existence and management of business processes in an organization is a suitable approach to materialize a value chain.

While the traditional conceptualizations of value chains and business processes may have evolved, their significance in the digital economy is undeniable. In this context, these concepts provide a foundational framework for the digitalization of organizational activities and the facilitation of seamless interoperability with business

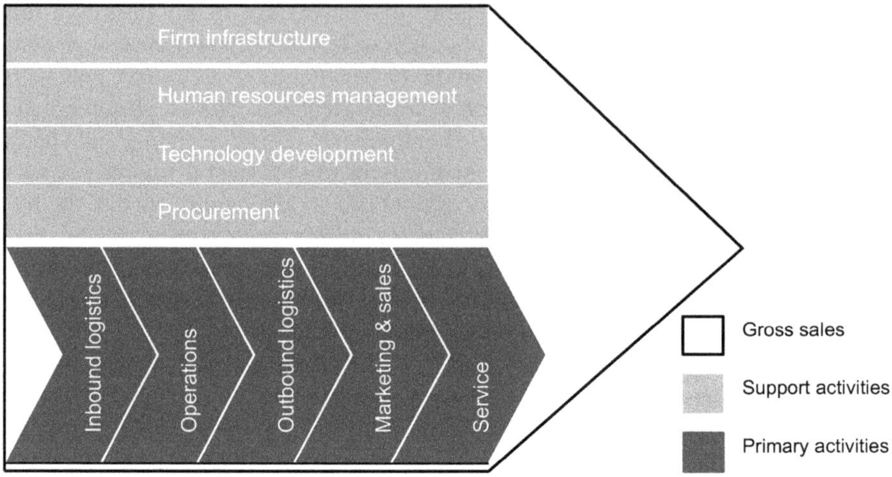

FIGURE 8.1 Value chain.

(Source: Adapted from Porter & Millar, 1985.)

partners and clients. The digital transformation of organizations is primarily achieved through software-based information systems. While information systems can theoretically exist in non-digital forms (e.g., traffic signs), the current and projected extents of digitalization in organizational activities render software-based solutions near-ubiquitous. Consequently, investigating the development methodologies of software products and services, their inherent business logic, and their interaction with existing information systems within the software landscape are becoming increasingly pertinent, as in the case of retailing value chains (Reinartz et al., 2019) and global value chains (Loonam & O'Regan, 2022).

Business processes can additionally serve as the primary source for structured data collection and processing. This facilitates the application of artificial intelligence algorithms for analysis, control, and predictive purposes, ultimately enabling real-time interventions within organizational operations.

This chapter explores the significance of business processes in the digital transformation of organizations and proposes a framework to facilitate the effective utilization of business process models in the development of enterprise systems within digital transformation initiatives.

Section 8.2 commences by exploring the relationship between the operational environments of organizations and the influence these environments exert on the configuration of their business processes. It then proceeds to propose target states for digitally transformed organizations. Section 8.3 explores the integration of business processes and software engineering. This integration facilitates the use of business process models as a foundation for software requirements. Subsequently, Section 8.4 proposes a framework that leverages business processes to support the digital transformation of organizations. Finally, Section 8.5 presents the key findings of this research and identifies potential avenues for future investigation.

8.2 DIGITAL TRANSFORMATION OF ORGANIZATIONS

As stated by Sony and Naik (2020), digital transformation is one of the major challenges of our times. For the past decades, organizations have been investigating digital technologies to enhance their products and their operations. These investigations frequently involve transformations of key business operations, affecting products and processes, as well as organizational structures and management concepts (Matt et al., 2015). Developing digital transformation requires using some kind of strategy, as explained by Hess et al. (2016) and Albukhitan (2020).

The digital transformation of an organization can occur at many distinct layers and can be driven by distinct goals. Porter and Heppelmann (2014) depict a scale of integration of physical and cyber-physical systems: at the lowest layer a product has only physical parts; at the following layer, smart products, the physical product has embedded some digital capabilities; in the next layer, smart connected products, smart products have a new digital part that is not computed at the edge, many times being a software application remotely accessible; at the fourth layer, product system, smart connected products cooperate in a digital system; at the final layer, system of systems, several product systems collaborate. This scale of integration exposes challenges that organizations need to address on their digital transformation path: What is the appropriated layer that their products need to address, knowing that the higher the address layer, the higher the complexity that needs to be tackled?

At the product system and at the system of systems layers, products are not standalone anymore and the extra connectivity and interoperability challenges play a major role on product development strategies and on product maintenance and support activities. For any of the five layers of the scale, business processes need to be:

- internally tailored, for example, how to order physical parts or how to define a development process for the embedded software resulting in a product with quality, in a timely way, with controlled costs, and fulfilling the requirements.
- externally compatible with the adequate business processes of external entities, like suppliers, technology partners, or clients.

One of the major cross-industry and world-wide movement on digital transformation is the Fourth Industrial Revolution, Industry 4.0 (I4.0), started from a German initiative materialized in 2011 (Vogel-Heuser & Hess, 2016) and followed by many other national initiatives around the globe (e.g., Produktion 2030 in Sweden, Industrial Internet Consortium in the USA, or Industria 4.0 in Italy). Previous industrial revolutions had, respectively, mechanization, electrification, and digitalization as disruptive technologies, while I4.0 is leveraged in real-time digital connections between all cyber-physical and digital entities participating in a business ecosystem. More recently, Industry 5.0 (I5.0) has appeared not as a technology-driven revolution but as a value-driven and human-focused revolution (Xu et al., 2021). I5.0 is currently conceptualized to leverage the unique creativity of human experts to collaborate with powerful, smart, and accurate machinery (Maddikunta et al., 2022). However, due, for example, to the enormous volumes of data, the velocity at which data is generated, the speed at which it must be analyzed, and the heterogeneity of data, humans may not be able to adequately cope with such challenges, making way for removing the humans in the loop. Following this perspective, Harari (2018) argues that humans will become irrelevant to the economy.

Since no new disruptive technology bases I5.0, the technological baseline of I4.0 is still valid and updated. In Lasi et al. (2014), two development directions for I4.0 are identified:

- An application-pull, triggered by short development cycles, flexibility, resource efficiency, among others.
- A technology-push, triggered by increasing mechanization and automation, miniaturization, and digitalization and networking.

The application-pull and technology-push directions induce a focus on business and information systems engineering driving technological and organizational implications, namely by moving the focus from product to service orientation, leading to approaches and technologies that allow the creation of business and technology-wise services. For that reason, the business process models and management emerge as an aggregating approach, because it has the potential to bring together digital and cyber-physical entities, either in a loosely or directive approach to business activities, and mapping business and technological contents. Even organizations that are not completely structured by business processes or are matrix-based must perform activities that can be framed within business process models.

As in I5.0, I4.0 also has humans as key players, even though their role is not always explicitly stated. Following an I4.0 agenda for the digital transformation of organizations, their products and services should include some features, like the autonomy and ability to allow individual smart connected products to take decisions, the use of open standards for the exchange of data and information, or the real-time digital representation of products and services during their complete lifecycle. These features are high-level requirements for the business contents of business processes and for software-based information systems.

To model an information system, the data–information–knowledge–wisdom (DIKW) hierarchy, depicted in Figure 8.2, attributed to Eliot (1934) or Ackoff (1989), should be used. According to Rowley (2007), data is defined as symbols representing the properties of objects, events, and their environment. Data is a product of observation. Information systems generate, store, retrieve, and process data. Information is inferred from data. Knowledge is what makes possible the transformation of information into instructions. Knowledge can be obtained by transmission, by instruction, or by extraction from experience.

Wisdom is defined as the ability to increase effectiveness and adds value requiring judgment. Additionally, intelligence is defined as the ability to increase efficiency.

A common concept in I4.0 is the Digital Twin (DT), initially named "Conceptual Ideal for Product Lifecycle Management" (Grieves, 2005). A DT is a digital informational construct about a physical system that can be an entity of its own. This digital representation would be a twin of the physical system and exhibits a real space, a virtual space, the link for data flow from real space to virtual space, the link for

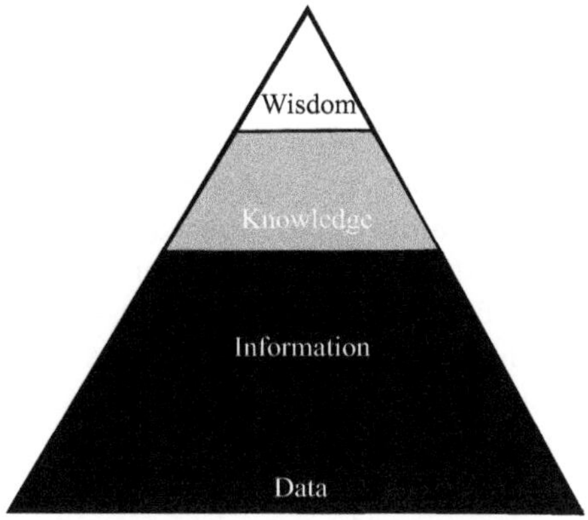

FIGURE 8.2 The DIKW pyramid.

(Source: Traditional visualization based on Van Meter, 2020.)

information flow from virtual space to real space, and virtual sub-spaces (Glaessgen & Stargel, 2012). DTs do not always contain all enumerated characteristics; for example, some organizations may consider a DT some digital model of a physical product, even without data and information connections. To allow some exchange of ideas from such diverse views, some maturity models, like the one proposed by Uhlenkamp et al. (2022), help on providing a common understanding of the contents and the scope of a given DT.

In Figure 8.3, we propose a mapping between the DIKW pyramid and the DT concept, covering the higher layers of the pyramid. Currently, edge devices have computational power and connectivity to perform knowledge and wisdom-related tasks.

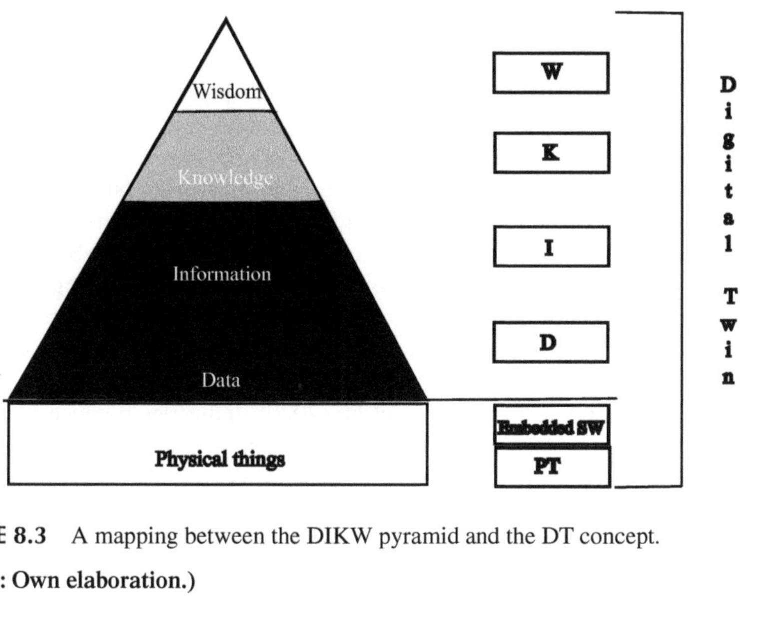

FIGURE 8.3 A mapping between the DIKW pyramid and the DT concept.

(Source: Own elaboration.)

For that reason, the place where computation occurs, that is, edge or cloud, is less relevant than the ability to perform it and synchronize both the physical and the virtual parts of the DT.

Within this perspective, physical things with embedded software, allowing computation and connectivity, can represent a digital twin of a smart connected product and the DT can exchange data, information, knowledge, and wisdom between its two parts. Additionally, the two parts do not need to be equal or to perform the same functions. For instance, a smart connected product can pass raw data from its physical part into its virtual part, which processes the data, and derive some knowledge and wisdom via an artificial intelligence algorithm (not present at the physical part), later passing only some instruction to the physical part.

Furthermore, DTs can map all the concepts of a digital transformation of an organization, like machines, production plans, production environment, weather conditions, products, and people.

8.3 BUSINESS PROCESS MODELS AS A FOUNDATION FOR SOFTWARE REQUIREMENTS

A business organization is an entity formed for the purpose of carrying on a commercial enterprise. Such an organization is predicated on systems of law governing contract and exchange, property rights, and incorporation (Encyclopaedia Britannica, 2010). Business organizations do not stand alone. They are part of constellations of organizations. Musiolik et al. (2020) study how innovation systems can be co-created with business partners based on resource constellations to better cope with the demand for the development of new technologies Operating complex and demanding business environments necessitate the strategic integration of internal business processes with those of suppliers, customers, and partners. MacDonald et al. (2022) state that "to address the prevalence and complexities of sustainable development challenges around the world, organizations in the business, government, and non-profit sectors are increasingly collaborating via multi-stakeholder partnerships".

Higgins (2005) suggests integration rules, for example, to standardize organizational processes and tools, establish a coherent organizational structure, or optimize the requirements and change management processes. Harmon (2019) highlights the importance of modeling, managing, integrating, and measuring business processes. Additionally, business organizations live inside a business environment where external and internal constraints shape internal structures and behaviors. Since business organizations are not freestanding entities, they must operate obeying constraints imposed by various forces, such as technological imperatives or the public interest. As pointed by Kay (1995), the concepts of industry and market are distinct and should not be confused when defining a business strategy, like when deriving a marketing strategy as mentioned by Möller et al. (2020). Industries produce services or products that markets consume. Thus, we place business organizations inside some industry or set of industries. The external environments surrounding an organization have a decisive impact on the structure and behavior of an organization. Campbell et al. (2002) state that external environments can be divided into two layers: macro- and microenvironment (see Figure 8.4). External macroenvironment may have impacts

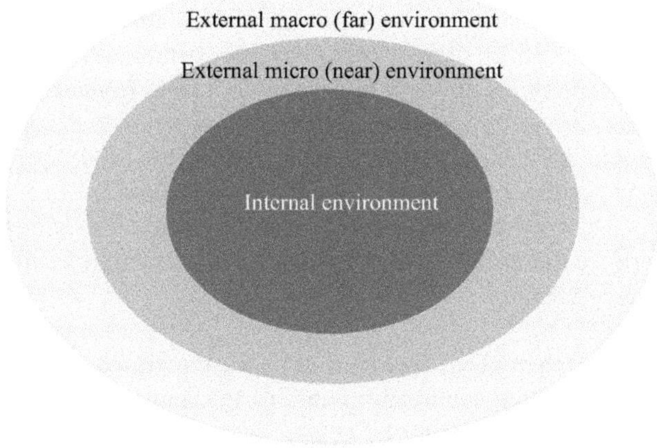

FIGURE 8.4 Environments of organizations.

(Source: Adapted from Campbell et al., 2002.)

from socio-demographic, technological, economical, legislative, natural environment, or even political factors. The macroenvironment is generally out of the influence of a single organization but may influence decisively the inner microenvironment. The industry, to which the organization belongs, and the market, where it sells its products or services, both compose the microenvironment of an organization.

Currently, organizations focus on the digital transformation of their internal environment. Many also focus on the external near environment due to the increasing need to cooperate with business partners and clients. Very few organizations focus on the external far environment because they can't influence it (e.g., laws, natural disasters, political decisions, or strikes in a far country), which sometimes leads to severe issues in the operation and even in the continuity of organizations, like the case of natural disasters in a chip-making country preventing the supply to its clients in other continents.

Since organizations can't influence the external far environment, many conclude that they should not care for it, but nothing prevents an organization from increasing the stock level in wintertime, when the roads where the suppliers have in-transit materials are with snow, compared to the summer situation, when the risk of the roads being blocked is lower.

Hence, all the three environments should be considered when digital transformation efforts occur and an explicit digital strategy for each one should be defined.

The ability to extract better results from a given market, or to foresee its behavior, is probably the most important success factor for an organization, either in the short-, mid-, or long-term time frames. To cope with external environments, organizations must shape themselves in such a way that they overcome external and internal constraints with success. Therefore, organizations draw strategies to achieve their business goals. Independently of the strategy model adopted, for example, Linear (Chandler, 1962), Adaptive (Hofer, 1973), or Blue Ocean (Kim & Mauborgne, 2005), an organization needs to change its internal environment to deal with the external business environments. Usually, in process-oriented organizations, a strategic management process materializes some form of business strategy.

To transform properly the strategy into operations, and to maintain a continuous link between both, organizations need a defined and structured approach. Among others, Davenport and Short (1990) and Kaplan and Norton (2008) propose a closed-loop management system, containing an integrated set of processes and tools that an organization can use to develop its strategy, translate it into operational actions, and monitor and improve the effectiveness of both. Five steps compose the management system of Kaplan and Norton:

1. Develop the strategy. Even prior to defining the strategy, managers need to define the mission (company's purpose), vision (aspiration for future results), and values (guide to the actions of the organization). Then, managers perform a strategic analysis, including political, economic, social, technological, environmental, and legal factors. Finally, they formulate the strategy by stating it and planning how the company proposes to achieve it.
2. Translate the strategy. In this step, managers need to translate the strategy into objectives and measures and clearly communicate it to all units and employees.
3. Plan operations, by improving important processes, developing sales plan, planning the needed capacity resources, and planning the budget.
4. Monitor and learn, by conducting strategy and operational reviews.
5. Test and adapt the strategy because assumptions underlying strategy are often flawed or obsolete.

The governance of the business processes after their deployment into the business reality should be resilient. Hence, a management system needs to be in operation, for instance by recurring to an organizational change management process. An information system must absorb the changes in the organization derived from the strategy and learned from the operations. For that, it is crucial that the project team understand the business strategy and its operationalization, to adapt properly the information and software systems and to design them recurring to technologies that backup the business strategy.

We follow the definition of business process from Rummler and Brache (1995):

> a series of steps designed to produce a product or service. Most processes are cross-functional, spanning the white space between the boxes on the organization chart. Some processes result in a product or service received by an organization's external customer. We call these primary processes. Other processes produce products that are invisible to the external customer but essential to the effective management of the business. We call these support processes.

Other definitions, such as from Davenport (1993), Hammer and Champy (1993), and Johansson et al. (1993), all have in common action-related elements (called either steps or activities) that start with some input and produce an output with value for the client. The focus on the client requirements is also a common characteristic of the definitions, making business processes an adequate model to cope with new challenges that occur daily during the normal business operation of an organization.

The ability to cope with incomplete and inconsistent requirements, and with their continuous change, from clients is bigger inside a process-oriented organization than in a department-oriented organization, because of the continuous focus of process-oriented organizations on clients' needs (normally expressed in the higher percentage of internal resources allocated to directly satisfying the client). Organizations should also be able to redefine their process landscapes according to new strategies or new clients' needs. A process landscape is not a static model, and therefore, the organization must regularly review it.

Due to all the previously mentioned factors, the design, the control, and the improvement of the business process landscape of an organization can only be efficient and useful if there exists a business process model that is defined, managed, and with quantifiable indicators.

8.4 LEVERAGING BUSINESS PROCESS MODELS FOR DIGITAL TRANSFORMATION INITIATIVES

Information systems are artifacts to study in their own right, independently of the characteristics of their users, the organizations in which they are employed, or the technologies used to implement them. Information systems have static and dynamic properties (Wand & Weber, 1990). Other terms for "information systems" include "enterprise application" or, for those with a long memory, "data processing" (Fowler, 2003).

Information systems are mediators between business frameworks and information technology and can be characterized by using in-depth system theoretical attributes. For example, the complexity of information systems is a significant system theoretical attribute. This complexity can be attributed to the fact that information systems work on a business, as well as on a technical, level (Thomas, 2006).

In a digital transformation project, information systems engineers should pay special attention to eventual conceptual dissonances between business processes and data and to the intrinsic complexity of the business logic (Fowler, 2003). Therefore, information systems should have a sufficient level of abstraction to embody concepts easily understood by humans, like invoices, production plans, or debit notes. Nevertheless, this high degree of abstraction must not be an excuse to not embody lower-level concepts, like the representation of an electrical signal occurring inside a business process. It is our opinion that information systems should be connected, for instance through a layered approach, to real-world events in real time. As of now, the huge volumes of data generated by the execution of business processes are stored, transformed, and later visualized by all sorts of business profiles to make proper business analysis and interventions. However, organizations should not end their digital transformation efforts when some fancy-looking cockpit charts made with state-of-the-art technologies are available but proceed into the realm of real-time data analysis and real-time intervention on business. This real-time approach needs the support of automation or artificial intelligence algorithms because it is not possible for humans to analyze such volumes of data, continually

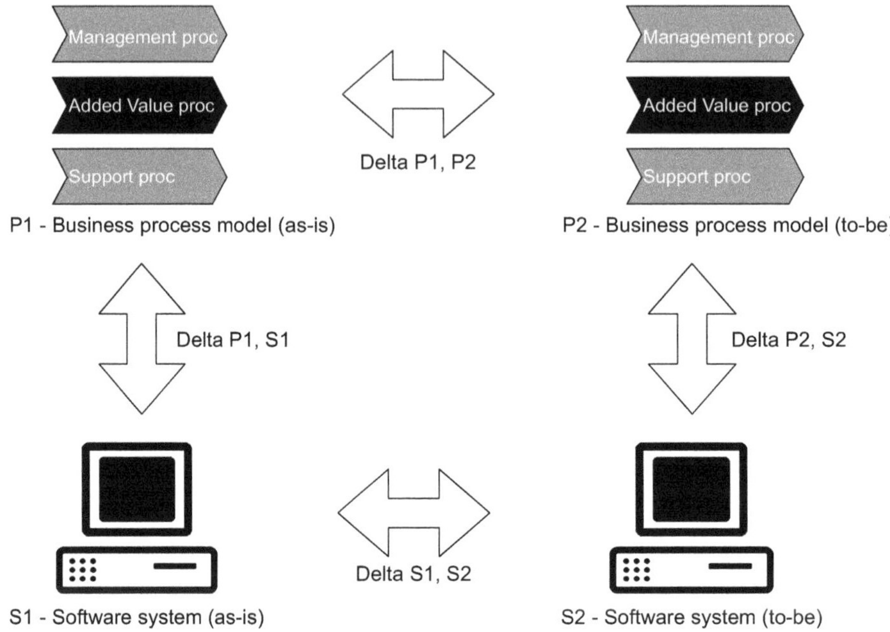

FIGURE 8.5 Differences between business process models and software systems, in "as-is" and "to-be" time frames.

(Source: Own elaboration.)

being streamed by sensors across all business processes, with a wide variety of formats and with unrelated contents. Several types of artificial intelligence algorithms can be chosen and combined according to a given purpose, such as analogizers, statistical, logic, genetic, and connectionists, but there is not yet an algorithm that can perform good enough for all purposes (Domingos, 2015), such as categorization or prediction. More recently, large language models, like Google Gemini or Microsoft CoPilot, are being applied to many business processes, and even new emergent abilities for such models are being discovered (Wei et al., 2022).

In Figure 8.5, we propose a framework component to allow checking the generic differences (Delta x, y) that may exist between real-world business processes (Px) and software-implemented business processes (Sy). S1 and P1 models represent the current moment in time ("as-is"), before the occurrence of some organizational or software change project, and S2 and P2 models represent a desired situation occurring in a future moment in time ("to-be").

Four sets of differences (represented by "Delta x, y" in Figure 8.5) may occur between real-world and software-implemented business processes:

1. Delta P1, P2: occurs when the current and the future real-world business processes have differences. This situation may occur when doing some business process improvement.

2. Delta S1, S2: occurs when the project team changes the software system from the current into the future situation, due to some requirement. Desirably, the transformation of S1 into S2 is supposed to occur simultaneously with the transformation of P1 into P2.
3. Delta P1, S1: relates with the differences between the real-world and software-implemented process models at the present time. For this delta, two main causes are possible:
   a. inside the current model of business processes (P1) exist a larger number of processes than those implemented in software (S1), configuring a normal situation in the sense that real-world business processes have a broader scope than their respective implementations in software systems. Nevertheless, this delta may also be caused by the software system not being supportive to the real-world processes;
   b. inside the current software model (S1) exist a larger number of processes than inside the present business process model (P1), which may occur due to an inadequately over-dimensioned software solution but also due to an inappropriate low level of use of the software system by the organization, like when using only some parts of an ERP system.
4. Delta P2, S2: in a future time, software-implemented business processes will be different from real-world business processes. Two main causes are possible:
   a. inside the future business process model (P2) exist more processes than inside the future software model (S2). This may occur in a normal situation, where an organization decides that not all new real-world processes (P2) are to be modeled in the new software system (S2) but may also occur because the new software model is not able to provide all the needed business functions, as in the case of a software system without quality;
   b. inside the future software model (S2) exist more business processes than those existing inside the future business process model (P2), this being a normal situation in the case of an organization knowing that the available software system should support future business expansion. It may also occur because the future software model (S2) is over-dimensioned.

From all the previously presented differences between real-world business models and software models, it is clear that a digital transformation project team should pay special attention to transformations from a present business model into a new one without the required transformation from the current into the new software model. This situation will lead to an execution of the new business model (P2) based on an outdated software model (P1), causing misunderstandings inside the organization due to the different behaviors in the real-world and the software models and creating serious problems with clients or suppliers.

In addition, a transformation of the present into the future software model can only be justified if the present software model is inadequate to support the present business model or if a technical change is happening.

As opposed to monolithic software applications, which are complex, difficult to scale, and continuously evolve, currently many software architectures (Bass et al., 2003) use an architectural style based on small, decentralized, independent, self-contained, and reusable software components, named microservices (Richardson, 2018). Microservices, and other architectural styles, like Enterprise Service Bus (Hohpe & Woolf, 2004), can be used to map parts of business processes as highly interconnected constellations of software components and reuse those components, at the time of operation, in several business processes at the same time. The mapping of business process models and software components is better performed when using a defined process, like that proposed by Duarte et al. (2012) and Santos et al. (2013).

Table 8.1 provides an overview of the three generations of information systems. First-generation systems focused on automating activities within individual departments of an organization. The second generation evolved to support operational activities across an entire organization, potentially spanning multiple sites. It is our contention that second-generation information systems currently represent the dominant paradigm, primarily driven by organizational priorities of integration within a single entity.

The third generation of information systems extends its focus beyond individual organizations, fostering linkages with suppliers, customers, and partner organizations. This approach, allowing inter-organizational BPM, strives to ensure ubiquitous availability of services to users while enabling business processes to execute seamlessly across multiple entities. While readily facilitated by contemporary technology, the extent to which organizations leverage this capability remains contingent upon their strategic mindsets, particularly regarding the potential substitution of human tasks with automated and artificial intelligence-based software components. Siau (2007) also provides scenarios for information system engineering. In this preview, two major scenarios are depicted for the third and subsequent generations of information systems:

TABLE 8.1
Evolution of the Focus of Information Systems Engineering

	First Generation	Second Generation	Third Generation
Scope	Single department	Single organization	Multiple organizations
Number of sites	Single site	One or few sites	Multiple sites and multiple countries
Organizational focus	Department level	Seamless integration within an organization	Seamless integration across multiple organizations
Technology focus	Automate business processes, business processes efficiency	Organization-wide closed loop	Integration of supply chain, e-commerce and e-market, mobility, and ubiquity
Information systems engineering methods and methodologies focus	Design and development focus (e.g., flowchart)	Support analysis, design, and development (e.g., ER, DFD)	Unification of modeling methods and methodologies (e.g., UML and UnifiedProcess)

Source: Adapted from Siau (2007).

- The Idealistic scenario, depicts major revolutions on the extent of automation in the analysis, design, implementation, and management of information systems. In addition, analysis and design of business applications can be done by end-users.
- In the Pragmatic scenario, components, commercial-off-the-shelf systems as well as metamodels, model-driven architecture (MDA) (Soley & OMG Staff Strategy Group, 2000), and service-oriented architecture (SOA) (Erl, 2005) will become fundamental areas. Domain ontologies will be further developed. It is also pointed out that some descendants of the Unified Process (Jacobson et al., 1999) will be used.

At the present time, there is a heightened interest in the Idealistic scenario. It is common practice in software development to use artificial intelligence chat bots to provide software engineers suggestions of code, and it is also common practice to use low-code or no-code software development tools that can be easily used by software engineers but also by business experts to develop software applications. It is noteworthy that business experts and end-users represent distinct stakeholder groups. Business experts possess specialized knowledge of business processes, structures, and activities. Consequently, they serve as a valuable source for both functional and non-functional requirements, particularly for backend systems (i.e., software not directly interacting with end-users). Conversely, end-users possess expertise in software application usage and provide valuable insights into non-functional requirements for frontend systems (i.e., software components directly interacting with humans). Notwithstanding this recognition, most software development activities continue to be undertaken by software engineers. Business experts and end-users mostly participate in hybrid teams together with software engineers or develop, on their own, first-generation information systems.

In Figure 8.6, the internal environment, including the internal business processes, and the external near environments, containing example business partner entities, are depicted. The external entities need to cooperate to perform a cross-border business process, like when a supplier receives a material order and delivers parts and, thus, interoperates with the inbound logistics business process. The integration of the internal business partners of an organization with the business processes of its business partners raises issues related to the following:

- Policies, that is, the willingness of all these organizations to share DIKW in a transparent, fluid, and real-time manner.
- Semantic interoperability, that is, how to create a common understanding of the exchanged DIKW allowing that all participants share the same semantics.
- Technological interoperability, that is, what are the most appropriate software architectures and technologies to materialize the cooperation between the business processes or where should the cross-organization business process, or its parts, be computed.

Organizations with open policies for digital interoperability have many advantages, like better and higher business opportunities, for example by showing their ability to

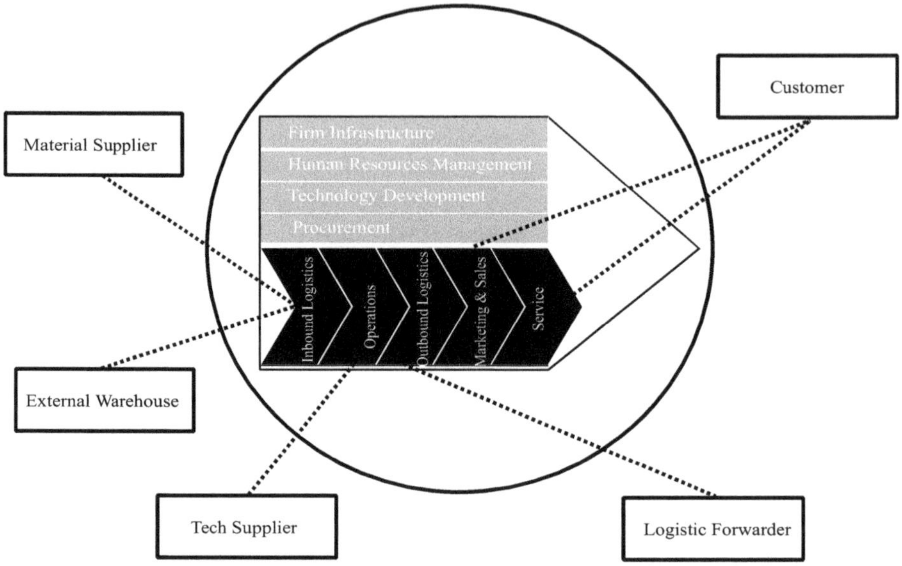

FIGURE 8.6 Interoperation at the internal and external near environments.

(Source: Own elaboration.)

digitally interface with big companies in real time in logistics or quality fields, and increased productivity, for instance due to the use of externally available master data instead of manually entering it. Many organizations still have a very strict use of DIKW in business processes that must interface with external organizations, like in inbound or outbound logistics, and not only on restricted business processes, like in the development of a new product or service.

Currently, the complexity in developing cross-organizational, or cooperating, information systems, together with the deep knowledge levels in areas like requirements, software architecture and design, software construction, test, development processes, or continuous development and delivery, requires that software projects, under these conditions, are handled by medium to large hybrid development teams. Additionally, if organizations want to map their far external environment, they need to digitally represent uncontrolled events, natural disasters, strikes, or changes in laws, by creating models for such events that must include the impacted internal and external near business processes. Despite modeling efforts being complex, organizations can start with simple models and later evolve them based on recorded own data or on purchased data, or add new algorithms to provide better predictions: in winter time, the estimation for the business process indicator related to on-time deliveries from suppliers can be lowered by some percentage, which affects other business processes from warehouse, production, or delivery. In terms of software architectures, several options exist to place the computation of the interoperable software-supported business processes:

- On an external partner, which normally occurs when an organization has a big supplier or customer that can impose such guidelines. In this way, powerful external partners dictate binding rules that prevent the organization from adapting their information systems, forcing it to develop new interfaces of its internal information system. This undesirable situation is amplified by the number of powerful partners the organization has.
- Internally, which, as opposite to the previous option, occurs when the organization is powerful enough to force it on the business partners, creating complexity outside its borders.
- Both internally and at each business partner, promoting that all participants change and adapt their information systems to allow interoperation between all business partners. Normally, open standards for semantic and technological interoperation are used. Nevertheless, the drawback is a point-to-point communication scenario, potentially forcing each participating organization to make changes to its internal information system each time a new business connection is established.
- Via an independent platform, where all organizations only need to create a single link to the platform, to connect to all their current and future business partners adhering to the platform. The platform is responsible for semantic interoperation and allows data and artificial intelligence algorithms to exist, improving end-to-end the business process.

As of now, the first three options are common. The fourth option, despite all the presented advantages and being better suited to improve end-to-end performance for the cross-border business process, is not common in the added-value and real-time business processes.

8.5 CONCLUSIONS AND FUTURE WORK

This chapter explores the significance of business processes in the digital transformation of organizations. It serves as a recommendation guide for organizations undertaking the development of software-based enterprise systems. Furthermore, this chapter proposes a framework to facilitate the effective utilization of business process models in the development of enterprise systems within digital transformation initiatives. The framework addresses four key areas:

- Interoperability across internal and external business processes.
- Discrepancies between real-world business processes and their digital representations.
- Insufficient data in digital transformation initiatives and the consideration of information, knowledge, and wisdom-related concerns.
- Establishing real-time connections between the physical and digital realms, where appropriate, for digitally represented business processes and relating it with Digital Twins.

Digital transformation of an organization is a multifaceted undertaking, best conceptualized as a portfolio encompassing numerous programs and projects. Business constraints, such as available financial resources or the pool of skilled collaborators, alongside the prevailing business context, significantly influence the digital transformation strategy. Nevertheless, organizations must integrate their digital maturity objectives into their overall business strategy and subsequently tailor their actions accordingly.

Digital transformation is a purpose that, in each business cycle, must be addressed and materialized in a portfolio. Organizations need to drive this transformation by planning maturity states, defining goals, setting up resources, and translating all into a coordinated action plan compatible with their business strategy.

Up to now, many digital transformation initiatives have been characterized by a partial and bottom-up approach. Organizations typically select isolated business pain points or areas with a high probability of success, marking the starting point for their digital transformation journey. While such organizations may claim to have embarked on digital transformation, this approach raises a crucial question: are they truly driving their own course and achieving the necessary levels of digital maturity, or are they simply following a fashionable trend with an uncertain endpoint? This dilemma represents one of the most significant strategic considerations for contemporary business leaders.

Data, information, knowledge, and wisdom are essential digital transformation areas that translate into concerns to be addressed by digital transformation programs, at the internal scope but also at the interoperation with business partners from the near external layer and at the far external layers of the business context of an organization. Platforms, independent from business partners, are a proper way to allow interoperation, in real time, with shared semantics, and with few technological preoccupations with present and future business partners, but they are not yet part of the digital offer available for organizations. Organizations must incorporate, among others, concerns of:

- Digitally representing all the “things”, either physical or digital, that participate in business processes (e.g., production stations, number of views of a social network, or the physical location of the purchased parts) and deal with the produced data in real time.
- Foreseeing future events that may affect, positively or negatively, the business.
- Promoting business process fluidity, that is, business processes should run smoothly and with minimal human intervention, in normal business contexts.

If organizations do not address the real-time characteristics of their business processes, the effectiveness and efficiency of all the efforts in business analytics and timely business interventions, either preventive or corrective, are hugely decreased. The time to have business managers looking at dozens of key process indicators in fancy-looking cockpit charts is over, due to data volumes, velocity of creation, and variety.

Future research on digital transformation will be focused on the shape of organizations and their surrounding contexts. Future organizations will be composed of intelligent and temporary aggregations of connected business functions, implemented in software. This vision, which is becoming increasingly feasible due to the digitalization of the economy, highlights the need for technological and semantic interoperability between the business processes of different organizations. Present-day software implementations of business functions (which are the basic components for the construction of business processes) are too static, require significant development times, and are dependent on human configuration. Therefore, it is important that software-implemented business functions be variable and self-reconfigurable. In addition to this internal intelligence, global intelligence and real-time behavior are also required in the interoperability with business processes of other organizations.

REFERENCES

Ackoff, R. L. (1989). From data to wisdom. *Journal of Applied Systems Analysis*, *16*(1), 3–9.

Albukhitan, S. (2020). Developing digital transformation strategy for manufacturing. *Procedia Computer Science*, *170*, 664–671.

Bass, L., Clements, P., & Kazman, R. (2003). *Software Architecture in Practice*. Addison-Wesley Professional.

Campbell, D., Stonehouse, G., & Houston, B. (2002). *Business Strategy: An introduction* (2nd Edition). Elsevier Butterworth-Heinemann.

Chandler, A. D. (1962). *Strategy and Structure: Chapters in the History of the American Enterprise*. Beard Books Inc.

Davenport, T., & Short, J. (1990). *The New Industrial Engineering: Information Technology and Business Process Redesign*. Center of Information Systems Research, Mass. Institute of Technology, Sloan School of Management.

Davenport, T. H. (1993). *Process Innovation: Reengineering Work through Information Technology*. Harvard Business School Press.

Denner, M.-S., Püschel, L. C., & Röglinger, M. (2018). How to exploit the digitalization potential of business processes. *Business & Information Systems Engineering*, *60*, 331–349.

Domingos, P. (2015). *The Master Algorithm: How the Quest for the Ultimate Learning Machine Will Remake Our World*. Basic Books.

Duarte, F. J., Machado, R. J., & Fernandes, J. M. (2012). BIM: A methodology to transform business processes into software systems. In S. Biffl, D. Winkler, & J. Bergsmann (Eds.), *Software Quality Days, SWQD 2012* (Vol. 94, pp. 39–58). Springer-Verlag.

EFQM. (2021). *EFQM Model* (2nd Edition). www.efqm.org

Eliot, T. S. (1934). *The Rock*. Faber & Faber.

Encyclopaedia Britannica. (2010, June). *Encyclopaedia Britannica*. Encyclopaedia Britannica. http://www.britannica.com/EBchecked/topic/86277/business-organization

Erl, T. (2005). *Service-oriented Architecture*. Prentice Hall.

Fowler, M. (2003). *Patterns of Enterprise Application Architecture*. Addison-Wesley Professional.

Glaessgen, E., & Stargel, D. (2012). The digital twin paradigm for future NASA and US Air Force vehicles. *53rd AIAA/ASME/ASCE/AHS/ASC Structures, Structural Dynamics and Materials Conference 20th AIAA/ASME/AHS Adaptive Structures Conference 14th AIAA, 1818*.

Grieves, M. W. (2005). Product lifecycle management: The new paradigm for enterprises. *International Journal of Product Development*, *2*(1–2), 71–84.

Hamel, G. (2000). *Leading the Revolution*, Harvard Business School Press.

Hammer, M., & Champy, J. (1993). *Reengineering the Corporation*. Harper Business.

Harari, Y. N. (2018). *21 Lessons for the 21st Century*. Random House.

Harmon, P. (2019). *Business Process Change: A Business Process Management Guide for Managers and Process Professionals*. Morgan Kaufmann.

Hess, T., Matt, C., Benlian, A., & Wiesböck, F. (2016). Options for formulating a digital transformation strategy. *MIS Quarterly Executive*, *15*(2).

Higgins, B. (2005). *Ground Rules for Managing Business Process Integration Projects. IBM Rational Edge*. http://www-128.ibm.com/developerworks/rational/library/jul05/higgins/index.html

Hofer, C. W. (1973). Some preliminary research on patterns of strategic behavior. *Academy of Management Proceedings*, *33*, 46–59.

Hohpe, G., & Woolf, B. (2004). *Enterprise Integration Patterns: Designing, Building, and Deploying Messaging Solutions*. Addison-Wesley Professional.

Imgrund, F., Fischer, M., Janiesch, C., & Winkelmann, A. (2018, March). Approaching Digitalization with Business Process Management. In *MKWI* (pp. 1725–1736). The work is in the proceedings of the conference MKWI 2018.

ISO. (2015). *ISO 9001:2015 Quality Management Systems*. https://www.iso.org/standard/62085.html

Jacobson, I., Booch, G., & Rumbaugh, J. (1999). The unified process. *IEEE Software*, *16*(3), 96.

Johansson, H., McHugh, P., Pendlebury, J. and Wheeler, W. (1993), *Business Process Reengineering: Break Point Strategies for Market Dominance*, John Wiley and Sons, Chichester.

Kaplan, R. S., & Norton, D. P. (2008). Mastering the management system. *Harvard Business Review*, *86*(1), 62.

Kay, J. A. (1995). *Why Firms Succeed*. Oxford University Press.

Kerpedzhiev, G. D., König, U. M., Röglinger, M., & Rosemann, M. (2021). An exploration into future business process management capabilities in view of digitalization: Results from a Delphi study. *Business & Information Systems Engineering*, *63*, 83–96.

Kim, W. C., & Mauborgne, R. (2005). *Blue Ocean Strategy: How to Create Uncontested Market Space and Make the Competition Irrelevant*. Harvard Business School Pr.

Kotarba, M. (2018). Digital transformation of business models. *Foundations of Management*, *10*(1), 123–142.

Lasi, H., Fettke, P., Kemper, H.-G., Feld, T., & Hoffmann, M. (2014). Industry 4.0. *Business & Information Systems Engineering*, *6*, 239–242.

Loonam, J., & O'Regan, N. (2022). Global value chains and digital platforms: Implications for strategy. *Strategic Change*, *31*(1), 161–177.

López, T., Riedler, T., Köhnen, H., & Fütterer, M. (2022). Digital value chain restructuring and labour process transformations in the fast-fashion sector: Evidence from the value chains of Zara & H&M. *Global Networks*, *22*(4), 684–700.

MacDonald, A., Clarke, A., & Huang, L. (2022). Multi-stakeholder partnerships for sustainability: Designing decision-making processes for partnership capacity. In *Business and the Ethical Implications of Technology* (pp. 103–120). Springer.

Maddikunta, P. K. R., Pham, Q.-V., Prabadevi, B., Deepa, N., Dev, K., Gadekallu, T. R., Ruby, R., & Liyanage, M. (2022). Industry 5.0: A survey on enabling technologies and potential applications. *Journal of Industrial Information Integration*, *26*, 100257.

Matt, C., Hess, T., & Benlian, A. (2015). Digital transformation strategies. *Business and Information Systems Engineering*, *57*, 339–343.

Miao, Z. (2021). Digital economy value chain: Concept, model structure, and mechanism. *Applied Economics*, *53*(37), 4342–4357.

Möller, K., Nenonen, S., & Storbacka, K. (2020). Networks, ecosystems, fields, market systems? Making sense of the business environment. *Industrial Marketing Management*, *90*, 380–399.

Musiolik, J., Markard, J., Hekkert, M., & Furrer, B. (2020). Creating innovation systems: How resource constellations affect the strategies of system builders. *Technological Forecasting and Social Change*, *153*, 119209.

OMG. (2011). Business process model and notation (BPMN), version 2.0. *OMG Document Number, Formal/2011-01-03*. http://www.omg.org/spec/BPMN/2.0/

OMG. (2017). *Unified Modeling Language (2.5.1)*. https://www.omg.org/spec/UML/2.5.1/About-UML

Osterwalder, A., Pigneur, Y., & Tucci, C. L. (2005). Clarifying business models: Origins, present and future of the concept. *Communications of AIS, 16*.

Porter, M. (1985). *Competitive Advantage: Creating and Sustaining Superior Performance*. Free Press.

Porter, M. E. (1980). *Competitive Strategy: Techniques for Analyzing Industries and Competitors*. Free Press.

Porter, M. E., & Heppelmann, J. E. (2014). How smart, connected products are transforming competition. *Harvard Business Review*, *92*(11), 64–88.

Porter, M. E., & Millar, V. E. (1985). How information gives you competitive advantage. *Harvard Business Review*, *63*(4), 149–160.

Porter, M. E., & Millar, V. E. (1999). How information gives you competitive advantage: The information revolution is transforming the nature of competition. *Knowledge and Special Libraries*, 85 85–103.

Reinartz, W., Wiegand, N., & Imschloss, M. (2019). The impact of digital transformation on the retailing value chain. *International Journal of Research in Marketing*, *36*(3), 350–366.

Richardson, C. (2018). *Microservices Patterns: With Examples in Java*. Simon and Schuster.

Rowley, J. (2007). The wisdom hierarchy: Representations of the DIKW hierarchy. *Journal of Information Science*, *33*(2), 163–180.

Rummler, G. A., & Brache, A. P. (1995). *Improving Performance: How to Manage the White Space on the Organization Chart*. Jossey-Bass San Francisco.

Santos, N., Duarte, F. J., Machado, R. J., & Fernandes, J. M. (2013). A transformation of business process models into software-executable models using MDA. In *Software Quality. Increasing Value in Software and Systems Development* (pp. 147–167). Springer.

Siau, K. (2007). The future of information systems engineering. *Requirements Engineering*, *12*(4), 199–201.

Smith, H., & Fingar, P. (2002). *Business Process Management: The Third Wave*. Meghan-Kiffer Press.

Soley, R. & OMG Staff Strategy Group. (2000). Model driven architecture. *OMG White Paper*, *308*(308), 5.

Sony, M., & Naik, S. (2020). Key ingredients for evaluating Industry 4.0 readiness for organizations: A literature review. *Benchmarking: An International Journal*, *27*(7), 2213–2232.

Thomas, O. (2006). Understanding the term reference model in information systems research: History, literature analysis and explanation. *Lecture Notes in Computer Science*, *3812*, 484.

Uhlenkamp, J.-F., Hauge, J. B., Broda, E., Lütjen, M., Freitag, M., & Thoben, K.-D. (2022). Digital twins: A maturity model for their classification and evaluation. *IEEE Access*, *10*, 69605–69635.

Van Meter, H. J. (2020). Revising the DIKW pyramid and the real relationship between data, information, knowledge, and wisdom. *Law, Technology and Humans*, *2*(2), 69–80.

Vogel-Heuser, B., & Hess, D. (2016). Guest editorial Industry 4.0 – Prerequisites and visions. *IEEE Transactions on Automation Science and Engineering*, *13*(2), 411–413.

Wand, Y., & Weber, R. (1990). An ontological model of an information system. *IEEE Transactions on Software Engineering*, *16*(11), 1282–1292.

Wei, J., Tay, Y., Bommasani, R., Raffel, C., Zoph, B., Borgeaud, S., Yogatama, D., Bosma, M., Zhou, D., Metzler, D., & others. (2022). Emergent abilities of large language models. *Transactions on Machine Learning*, 08/2022. https://arxiv.org/pdf/2206.07682.pdf

Xu, X., Lu, Y., Vogel-Heuser, B., & Wang, L. (2021). Industry 4.0 and Industry 5.0 – Inception, conception and perception. *Journal of Manufacturing Systems*, *61*, 530–535. https://doi.org/10.1016/j.jmsy.2021.10.006

Index

Pages in **bold** refer to tables

For Product Safety Concerns and Information please contact our EU
representative GPSR@taylorandfrancis.com
Taylor & Francis Verlag GmbH, Kaufingerstraße 24, 80331 München, Germany

www.ingramcontent.com/pod-product-compliance
Ingram Content Group UK Ltd.
Pitfield, Milton Keynes, MK11 3LW, UK
UKHW020206200926
475702UK00004B/641

* 9 7 8 1 0 3 2 6 1 0 0 0 9 *